T0327713

Annals of Mathematics Studies

Number 83

# AUTOMORPHIC FORMS ON ADELE GROUPS

BY

STEPHEN S. GELBART

PRINCETON UNIVERSITY PRESS
PRINCETON, NEW JERSEY

Printed in the United States of America

LIBRARY OF CONGRESS CATALOGING IN PUBLICATION DATA

Gelbart, S.
Automorphic forms on Adele groups.
(Annals of mathematics studies; no. 83)
"Expanded from notes mimeographed at Cornell in May of 1972 and
entitled Automorphic forms and representations of Adele groups."
Bibliography: p.
Includes index.
1. Representations of groups. 2. Automorphic forms.
3. Linear algebraic groups.
I. Title. II. Series.
QA171.G39 1975 512'.22 74-23388
ISBN 0-691-08156-5

3 5 7 9 10 8 6 4 2

To my father,
Abe Gelbart

## PREFACE

Sections 1 through 7 of these Notes are based on lectures I gave at Cornell University in the Spring of 1972. They are expanded from Notes mimeographed at Cornell in May of 1972 and entitled *Automorphic Forms and Representations of Adele Groups*. I am grateful to E. M. Stein for suggesting that I expand those Notes for publication by the Princeton University Press and that I incorporate into them the material of Sections 8 through 10. These last three sections are based on lectures I gave at the Institute for Advanced Study, Princeton, in the Spring of 1973. I am indebted to the Institute for its hospitality as well as for the atmosphere it created for serious work.

The subject matter of these Notes is the interplay between the theory of automorphic forms and group representations. One goal is to interpret some recent developments in this area, most significantly the theory of Jacquet-Langlands, working out, whenever possible, explicit consequences and connections with the classical theory. Another goal is to collect as much information as possible concerning the decomposition of $L^2(GL(2,Q)\backslash GL(2,A(Q))$. Although each particular section has its own introduction describing the material covered I would like to add the following orienting remarks to this Preface.

Sections 1 through 5 are preliminary in nature and their purpose is to spell out the explicit relations between classical cusp forms and certain irreducible constituents of $L^2(GL(2,Q)\backslash GL(2,A(Q))$. Here I collect only those facts from representation theory and the classical theory of forms which are crucial to the sequel. Parts of these sections are either new or part of the subject's "folklore." References to the existing literature are to be found in the "Notes and References" at the end of each section and individual acknowledgements are made whenever possible.

Sections 6 through 10 deal with Jacquet-Langland's theory and some important questions related to it. Section 8 describes the continuous spectrum of $L^2(GL(2,\mathbb{Q})\backslash GL(2,\mathbb{A}))$ and is perhaps the least self-contained. The remaining sections, including Section 9 on the trace formula, concern the discrete spectrum. I have included a complete proof of the trace formula for GL(2) primarily because the ideas involved here are still not well known. I also wanted there to be no doubt in the reader's mind that the proof of Jacquet-Langlands' Theorem 10.5 is now complete. In writing Section 9 I have followed J. G. Arthur's as yet unpublished manuscript on the trace formula for rank one groups and I wish to thank him for allowing me to do so.

Scattered throughout these Notes are some new results and proofs which I have not described elsewhere. I am indebted to my colleagues at Cornell, in particular K. S. Brown, W. H. J. Fuchs, A. W. Knapp, S. Lichtenbaum, O. S. Rothaus, R. Stanton, H. C. Wang, and W. C. Waterhouse, for help and encouragement, and to J. G. Arthur, P. Cartier, W. Casselman, R. Howe, R. Hotta, M. Karel, R. P. Langlands, R. Parthasarathy, P. J. Sally, Jr., and T. Shintani, for helpful conversations and correspondence related to the results described here. I especially wish to thank R. P. Langlands for much valuable information and inspiration.

The first typing of these Notes was done at Cornell by Esther Monroe, Dolores Pendell and Ruth Hymes. Their unusual efficiency and expertise was greatly appreciated.

ITHACA
DECEMBER 1973

CONTENTS

Automorphic Forms on

Adele Groups

## §1. THE CLASSICAL THEORY

This section describes various aspects of Hecke's theory of Dirichlet series attached to cusp forms and some recent refinements of it due to Weil and Atkin-Lehner. These results from the classical theory of automorphic forms play a crucial role in the modern theory. Since we include them primarily to provide a convenient classical reference for our discussion of Jacquet-Langlands' theory no attempt at completeness is made.

### A. Elementary Notions

Throughout this section we shall be dealing with *non* co-compact arithmetic subgroups of $SL(2,\mathbf{R})$. (The case of *compact* fundamental domain will be considered in Section 10.) In fact, $\Gamma$ will usually denote a congruence subgroup, i.e., a subgroup of $SL(2,\mathbf{Z})$ which contains the *homogeneous principal congruence subgroup*

$$\Gamma(N) = \left\{ \begin{bmatrix} a & b \\ c & d \end{bmatrix} \in SL(2,\mathbf{Z}) : \begin{bmatrix} a & b \\ c & d \end{bmatrix} \equiv \begin{bmatrix} 1 & 0 \\ 0 & 1 \end{bmatrix} (\text{mod } N) \right\}$$

for some positive integer $N$. Important examples are $SL(2,\mathbf{Z})$ (the "full congruence subgroup," or "congruence subgroup of level 1") and "Hecke's subgroup"

$$\Gamma_0(N) = \left\{ \begin{bmatrix} a & b \\ c & d \end{bmatrix} \in SL(2,\mathbf{Z}) : c \equiv 0 (\text{mod } N) \right\} .$$

By $GL^+(2,\mathbf{R})$ we shall denote the group of real $2 \times 2$ matrices with positive determinant. If $g = \begin{bmatrix} a & b \\ c & d \end{bmatrix}$ belongs to $GL^+(2,\mathbf{R})$, $z$ to $\{\mathrm{Im}(z) > 0\}$, and $k$ is a positive integer, we set

$$g(z) = \frac{az+b}{cz+d} , \tag{1.1}$$

(1.2) $$j(g,z) = (cz+d)(\det g)^{-1/2},$$

and

(1.3) $$f|_{[g]_k}(z) = f(gz)\,j(g,z)^{-k}.$$

This last formula defines an operator on the space of all complex-valued functions $f(z)$, $z \in \{\mathrm{Im}(z) > 0\}$.

Two points $z_1, z_2$ will be called *equivalent under* $\Gamma$ (or $\Gamma$-*equivalent*) if $\gamma z_1 = z_2$ for some $\gamma \in \Gamma$. A subset $F$ of $\{\mathrm{Im}(z) > 0\}$ is a *fundamental domain for* $\Gamma$ if $F$ is a connected open subset of $\{\mathrm{Im}(z) > 0\}$ with the property that no two points of $F$ are $\Gamma$-equivalent and *each* point of $\{\mathrm{Im}(z) > 0\}$ *is* $\Gamma$-equivalent to some point of the closure of $F$.

A point $s$ in $\mathbb{R} \cup \{\infty\}$ is a *cusp of* $\Gamma$ if there exists a *parabolic* element of $\Gamma$ fixing $s$. If $H^*$ denotes the union of $\{\mathrm{Im}(z) > 0\}$ and the cusps of $\Gamma$ then $\Gamma$ also acts on $H^*$; the resulting quotient space possesses a natural (Hausdorff) topology and a complex structure such that $\Gamma\backslash H^*$ is a *compact Riemann surface*.

The cusps we shall consider may be taken as various rational points on the real axis and $\infty$. Most authors denote the cusp at $\infty$ by $i\infty$ to emphasize that as $z = x+iy$ approaches the cusp in $F$, $x$ is bounded, and $y \to \infty$.

In general, if $\Gamma$ is an arbitrary discrete subgroup of $SL(2,\mathbb{R})$, $\Gamma$ is called a *Fuchsian group of the first kind* if $\Gamma\backslash H^*$ is compact. All Fuchsian groups, and $\Gamma_0(N)$ in particular, have (at most) a finite number of $\Gamma$-inequivalent cusps.

The following definition is valid for $\Gamma$ an arbitrary Fuchsian group of the first kind.

DEFINITION 1.1. A complex-valued function $f(z)$ is called a $\Gamma$-*automorphic form of weight* $k$ ( or an *automorphic form of weight* $k$ *for* $\Gamma$) if it is defined in $\{\mathrm{Im}(z) > 0\}$ and satisfies the following conditions:

(i) $f|[\gamma]_k \equiv f$, i.e.

$$f\left(\frac{az+b}{cz+d}\right) = (cz+d)^k f(z)$$

for all $\gamma = \begin{bmatrix} a & b \\ c & d \end{bmatrix} \in \Gamma$: this is the "automorphy condition" for $f$;

(ii) $f$ is holomorphic in $\{\mathrm{Im}(z) > 0\}$; and

(iii) $f$ is holomorphic at every cusp of $\Gamma$.

The space of such functions will be denoted $M_k(\Gamma)$.

For congruence subgroups, elements of $M_k(\Gamma)$ are often called *modular forms* (or *modular forms of level* $N$ if $\Gamma = \Gamma(N)$).

If $\psi$ is a character modulo $N$ (a character of $(\mathbb{Z}/N\mathbb{Z})^x$ extended in the obvious way to $\mathbb{Z}$), and $f(z)$ satisfies (in place of (i) above)

$$f\left(\frac{az+b}{cz+d}\right) = \psi(a)^{-1}(cz+d)^k f(z) ,$$

for all $\gamma \in \Gamma_0(N)$, then $f(z)$ is an automorphic form of weight $k$ *and character* $\psi$. The space of all such $f(z)$ is denoted $M_k(N,\psi)$.

REMARK 1.2. (*Concerning the notion of regularity at a cusp.*) Suppose $s$ is a cusp of $\Gamma$ and $\sigma$ in $SL(2,\mathbb{Z})$ maps $\infty$ to $s$. (Such $\sigma$ exist because $\infty$ is $SL(2,\mathbb{Z})$-equivalent to all rational points on the real axis.) From Condition (i) of Definition 1.1 it follows that $f|[\sigma]_k$ is invariant under $\rho = \sigma^{-1}\gamma\sigma$, $\gamma \in \Gamma_s$, if

$$\Gamma_s = \{\gamma \in \Gamma : \gamma(s) = s\} .$$

Indeed,

$$\begin{aligned} f|[\sigma]_k|[\rho]_k(z) &= f|[\sigma\rho]_k(z) = f|[\gamma\sigma]_k(z) \\ &= f(\gamma\sigma(z))\, j(\gamma\sigma,z)^{-k} = f(\gamma(\sigma(z))\, j(\gamma,\sigma(z))^{-k}\, j(\sigma,z)^{-k} \\ &= f|[\gamma]_k(\sigma z)\, j(\sigma,z)^{-k} = f(\sigma z)\, j(\sigma,z)^{-k} = f|[\sigma]_k(z) . \end{aligned}$$

But each $\rho$ in $\sigma^{-1}\Gamma_s\sigma$ is translation by some $h_1$, since $\sigma^{-1}\Gamma_s\sigma$ fixes $\infty$. Thus if $h$ denotes the smallest positive such $h_1$, *and* $k$ *is even*, we at least have

$$f|_{[\sigma]_k}(z+h) = f|_{[\sigma]_k}(z),$$

from which it follows that the function

$$\hat{f}_s(\zeta) = f|_{[\sigma]_k}(z), \tag{1.4}$$

with

$$\zeta = e^{2\pi iz/h} \quad \text{(the local uniformizing variable at } s) \tag{1.5}$$

is well defined in $|\zeta| < 1$, and holomorphic in the punctured disc (by Condition (ii)). *The precise meaning of Condition (iii) then is that* $\hat{f}_s(\zeta)$ *is regular at* $\zeta = 0$ *for every cusp* $s$.

This notion of regularity at a cusp $s$ (as well as the notion of "local uniformizing variable" and "width" $h$ of the cusp at $s$) *is independent of the choice of* $\sigma$. It should also be clear that $f$ is regular at the cusps of $\Gamma$ if and only if $f|_{[\sigma]_k}$ is regular at $\infty$ for all $\sigma \in SL(2,\mathbf{Z})$. Thus (for congruence subgroups) Conditions (ii) *and* (iii) of Definition 1.1 are independent of the group we regard $f$ as being a form for. In particular, if $f$ is a form for $\Gamma$, and $\Gamma' \subset \Gamma$, then $f$ is also a form for $\Gamma'$.

*When* $k$ *is odd*, some additional remarks are in order. For example, if $k$ is odd, and $\begin{bmatrix} -1 & 0 \\ 0 & -1 \end{bmatrix} \in \Gamma$, then $M_k(\Gamma)$ is zero. (Indeed Condition (i) of Definition 1.1 implies that $f(z) = -f(z)$. Thus we assume that $-1 \notin \Gamma$. If $\sigma^{-1}\Gamma_s\sigma$ is generated by $\begin{bmatrix} -1 & -h \\ 0 & -1 \end{bmatrix}$, then $f|_{[\sigma]_k}(z+h) = -f(z)$, so the variable $\zeta$ used to define regularity at $s$ should be $e^{\pi iz/h}$ (and not $e^{2\pi iz/h}$).

REMARK 1.3. (*Concerning the notion of Fourier expansion at a cusp.*) Suppose $f \in M_k(\Gamma)$. The fact that $f$ is regular at the cusp $s$ means that $\hat{f}_s(\zeta)$ has a Taylor series at $0$, which series in turn induces the expansion

$$f|_{[\sigma]_k}(z) = \sum_{n=0}^{\infty} a_n e^{2\pi inz/h}. \tag{1.6}$$

The series (1.6) converges absolutely and uniformly on compact subsets and is called the *Fourier expansion of* f *at the cusp* s, its coefficients $a_n$ the *Fourier coefficients of* f *at* s. As we shall see, this notion of "Fourier expansion at a cusp" is crucial to the entire theory of automorphic forms.

If $\Gamma$ is Hecke's subgroup $\Gamma_0(N)$, the Fourier expansion at $\infty$ of any f in $M_k(\Gamma)$ will be of the form

$$f(z) = \sum_{n=0}^{\infty} a_n e^{2\pi inz} . \tag{1.7}$$

Henceforth we shall be dealing almost exclusively with Fourier expansions of this type.

DEFINITION 1.4. A $\Gamma$-automorphic form is a *cusp form* if it vanishes at every cusp of $\Gamma$, i.e., its zeroth Fourier coefficient at each cusp is zero.

The space of $\Gamma$-cusp forms of weight k will be denoted $S_k(\Gamma)$. In general, from (1.6) it follows that the zeroth Fourier coefficient of f at s is given by

$$a_0(s) = \int_0^1 f|[\sigma]_k(hx+iy)dx . \tag{1.8}$$

In particular, if $\Gamma = \Gamma_0(N)$,

$$a_0 = a_0(\infty) = \int_0^1 f(x+iy)dx ,$$

where $y > 0$ is arbitrary.

LEMMA 1.5. *Suppose* $f \in M_k(\Gamma)$. *Then* $f \in S_k(\Gamma)$ *if and only if*

$$y^{k/2} |f(x+iy)| < M \tag{1.9}$$

*for some constant* M *independent of* x.

*Proof.* (For convenience, assume $k$ even.) The function $g(z) = (\mathrm{Im}(z))^{k/2} f(z)$ is easily seen to be $\Gamma$-invariant. What has to be shown is that $f(z)$ vanishes at each cusp $s$ of $\Gamma$ if and only if $|g(z)| < M$. But

$$f|[\sigma]_k(z) = \hat{f}_s(\zeta) ,$$

with $\zeta = e^{2\pi z/h}$ (as in (1.4) and (1.5)), so

$$g(\sigma(z)) = \mathrm{Im}(z)^{k/2} \hat{f}_s(\zeta)$$

(since $\mathrm{Im}(\sigma(z)) = \mathrm{Im}(z)|j(\sigma,z)|^{-2}$). Therefore, if $f \in S_k(\Gamma)$, $g(w) \to 0$ as $w \to s$ (with respect to the topology of $H^*$). This means that $g$ is a continuous function on the *compact* space $\Gamma\backslash H^*$ and hence $|g(z)| < M$. Conversely, if $|g(z)| < M$, $\hat{f}_s(\zeta)$ must be holomorphic at $\zeta = 0$, and in fact it must vanish there. □

COROLLARY 1.6. *If* $f(z) = \sum_{n=1}^{\infty} a_n e^{2\pi i n z} \in S_k(\Gamma)$, *then*

$$a_n = 0(n^{k/2}) . \tag{1.10}$$

*Proof.* Since $\hat{f}_\infty(\zeta) = \sum_{n=1}^{\infty} a_n \zeta^n$,

$$a_n = \frac{1}{2\pi i} \int \frac{\hat{f}_\infty(\zeta)}{\zeta^{n+1}} d\zeta ,$$

where the integral is taken over a small circle of radius $r$ about the origin. But by the lemma, $|\hat{f}_\infty(\zeta)| \leq M y^{-k/2}$, so taking $y = 1/n$ yields the desired conclusion. □

REMARK 1.7. (Petersson inner product.) If $f$ and $g$ belong to $S_k(\Gamma)$, we define their scalar product by

$$(f,g) = (f,g)_{k,\Gamma} = \iint_F f(z)\overline{g(z)}\, y^k \frac{dxdy}{y^2} , \tag{1.11}$$

where $z = x+iy$ and $F$ is any fundamental domain for $\Gamma$. This definition is independent of the choice of $F$ since the "hyperbolic" measure

$$\frac{dxdy}{y^2}$$

(like $f(z)\overline{g(z)}y^k$) is invariant under $\Gamma$. As for convergence of the double integral in (1.11), this follows from the fact that

$$\text{(1.12)} \qquad f(z) \text{ in } S_k(\Gamma) \text{ is } 0(e^{-2\pi y/h}) \quad (y \to \infty)$$

(cf. (1.6)).

Equipped with this inner product, $S_k(\Gamma)$ is a finite-dimensional Hilbert space.

B. Examples

(i) *Ramanujan's Function.* Let $\Delta(z)$ denote the function defined in $\{\mathrm{Im}(z) > 0\}$ by

$$e^{2\pi iz} \prod_{n=1}^{\infty} (1-e^{2\pi inz})^{24} .$$

This function was investigated in 1916 by Ramanujan. It is a cusp form of weight 12 for the full modular group, its Fourier expansion (at $\infty$) is

$$\Delta(z) = \sum_{n=1}^{\infty} \tau(n)e^{2\pi inz} ,$$

and Ramanujan conjectured

$$\tau(n) = 0(n^{11/2+\varepsilon}), \varepsilon > 0 .$$

(ii) *Poincaré Series*. Suppose $\Gamma$ is a subgroup of finite index in $SL(2,Z)$, and $\Gamma_0$ is the infinite cyclic subgroup of translations in $\Gamma$ generated by the least translation $z \to z+q$.

Then for any positive integer k, and non-negative integer $\nu$, the *Poincaré series of weight* k *and character* $\nu$ is defined by the series

(1.13) $$\phi_\nu(z) = \sum_\gamma e^{2\pi i\nu \frac{\gamma(z)}{q}} (cz+d)^{-2k} ,$$

the summation extending over a set of representatives $\gamma = \begin{bmatrix} a & b \\ c & d \end{bmatrix}$ of $\Gamma_0$ in $\Gamma$. This Poincaré series converges absolutely uniformly on compact subsets of $\{\mathrm{Im}(z) > 0\}$ and describes there a $\Gamma$-automorphic form of weight 2k. In fact, for $\nu \geq 1$, $\phi_\nu(z)$ is a cusp form. Furthermore, using the Petersson inner product, it can be shown that *every* f *in* $S_{2k}(\Gamma)$ *is a linear combination of the Poincaré series* $\phi_\nu(z)$, $\nu \geq 1$.

(iii) *Analytic Eisenstein Series*. The Poincare series $\phi_0(z)$ is not a cusp form. Indeed, for $\Gamma = SL(2,Z)$,

$$\phi_0(z) = \sum_{(c,d)} \frac{1}{(cz+d)^{2k}} ,$$

the summation extending over all pairs of integers $(c,d) \neq (0,0)$. The function

(1.14) $$E_k(z) = \frac{1}{2\zeta(2k)} \phi_0(z) \quad (\zeta(s) = \Sigma n^{-s})$$

is a modular form of weight 2k for $\Gamma$, called the (normalized) *Eisenstein series*; if $B_n$ denotes the n-th Bernoulli number then

(1.15) $$E_k(z) = 1 + \frac{(-1)^k 4k}{B_k} \sum_{n=1}^{\infty} \sigma_{2k-1}(n) e^{2\pi i n z}$$

describes the Fourier expansion of $E_k(z)$ at $\infty$. (Here $\sigma_r(n) = \sum_{d|n} d^r$.)

(iv) *Theta Series.* We give only the simplest example. Suppose $A$ is a positive definite symmetric *even* integral $r \times r$ matrix. (*Even* integral means that $a_{ij}$ and $\frac{1}{2}a_{ii}$ are all integral.) Let us also suppose that $A^{-1}$ is even integral. (This last assumption actually implies that $r$ is divisible by 8.)

To $A = (a_{ij})$ we associate the quadratic form $Q(X) = \Sigma a_{ij}x_{ij} = {}^tXAX$, and a function

(1.16) $$\theta_Q(z) = \sum_M e^{\pi i z Q(M)}$$

where $M$ runs through all integral vectors $M = \begin{pmatrix} m_1 \\ \vdots \\ m_r \end{pmatrix}$. The series in (1.16) defines the *theta function associated to the quadratic form* $Q$. Using the Poisson summation formula one proves that $\theta_Q(z)$ *is a modular form of weight* $r/4$ *for the full modular group.*

The arithmetic significance of this example is the following. Suppose $r = 8k$ and $r(n)$ is the number of ways of expressing $2n$ as $Q(M)$. Clearly

$$\theta_Q(z) = \sum_{n=0}^{\infty} r(n) e^{2\pi i n z} .$$

But $\theta_Q(z) - E_k(z)$ is easily seen to be a *cusp* form of weight $2k$, so from (1.10) and (1.15) we conclude

$$r(n,Q) = \frac{4k}{B_k} \sigma_{2k-1}(n) + 0(n^k) ,$$

an estimate which becomes an *exact formula* for $r(n,Q)$ when $k$ is such that $S_{2k}(\Gamma)$ is empty (e.g. $k = 2,4$).

If we drop the assumption that $A^{-1}$ is even integral (so $r$ is not necessarily even let alone divisible by 8!) then the theta-series associated to the quadratic form $Q(X) = {}^tXAX$ is still an automorphic form but no longer for the full congruence group and no longer necessarily of integral weight. The classical example here is

$$\theta(z) = \sum_{n=-\infty}^{\infty} e^{2\pi i n^2 z} ,$$

an automorphic form of "half-integral weight" for $\Gamma_0(4)$. (For a theory of forms of half-integral weight see [Shimura 2].)

Further relations between theta-series and automorphic forms, especially from the modern point of view, will be discussed in Section 10.

## C. Hecke's Theory

For our purposes it will be sufficient to sketch Hecke's theory for cusp forms *on the special congruence subgroups* $\Gamma_0(N)$.

Roughly speaking, Hecke's theory associates to each $f(z) = \Sigma a_n e^{2\pi i n z}$ in $S_k(N,\psi)$ a Dirichlet series $D(s,f) = \sum_1^\infty a_n n^{-s}$ which is shown to possess an analytic continuation and simple functional equation. The theory also establishes the existence of a basis for $S_k(N,\psi)$ consisting of functions whose Fourier coefficients satisfy certain multiplicative properties of number theoretic interest.

We start by recalling the Hecke operators.

For each prime p we consider the double coset

$$\Gamma_0(N)\begin{bmatrix}1 & 0\\ 0 & p\end{bmatrix}\Gamma_0(N) = \bigcup_j \Gamma_0(N)\gamma_j = \bigcup_{(a,N)=1}\ \bigcup_{b=0}^{d-1} \Gamma_0(N)\,\sigma_a\begin{bmatrix}a & b\\ 0 & d\end{bmatrix}$$

$$ad = p, \quad a > 0$$

where $\sigma_a = \begin{bmatrix}a' & b'\\ 0 & d'\end{bmatrix}$ in SL(2,Z) is chosen congruent to $\begin{bmatrix}a & 0\\ 0 & a^{-1}\end{bmatrix}$ modulo N. The p-th *Hecke operator* T(p) is the operator defined on $S_k(N,\psi)$ through the natural action of this double coset. More precisely, from (1.3), we put

$$\begin{aligned} T_k(p)f &= f|_{[\Gamma_0(N)\begin{bmatrix}1 & 0\\ 0 & p\end{bmatrix}\Gamma_0(N)]_k} \\ &= p^{k/2-1}\sum_j f|_{[\gamma_j]_k} \\ &= p^{k-1}\sum_{\substack{a>0\\ ad=p}}\ \sum_{b=0}^{d-1} \psi(a) f\left(\frac{az+b}{d}\right) d^{-k} \end{aligned}$$

(On functions, $\sigma_a$ operates as multiplication by $\psi(a)$!).

Now suppose $(p,N) = 1$ and $\psi$ is trivial. Then for all $f, g \in S_k(N,\psi)$,

$$(T(p)f,g) = (f,T(p)g) \;, \tag{1.19}$$

i.e. $T(p)$ is hermitian with respect to the Petersson inner product.

In general, if $\psi$ is arbitrary, and $p$ is still relatively prime to $N$,

$$(T(p)f,g) = \psi(p)(f,T(p)g) \;.$$

Moreover, the algebra of operators generated by these $T(p)((p,N)=1)$ is a *commutative* algebra of normal operators on $S_k(N,\psi)$ and consequently there exists a basis for $S_k(N,\psi)$ consisting of functions which are simultaneous eigenfunctions for all these $T(p)$. The significance of such eigenfunctions is the following.

Suppose $f(z)$ belongs to $S_k(N,\psi)$ and its Fourier expansion (at $\infty$) is

$$f(z) = \sum_{n=1}^{\infty} a(n)e^{2\pi inz} \;.$$

Then if $g(z) = (T(p)f)(z) = \sum_{n=1}^{\infty} a'(n)e^{2\pi inz}$,

$$a'(n) = \sum_{d|(n,p)} \psi(d)d^{k-1}a\left(\frac{np}{d^2}\right) = a(np) + \psi(p)p^{k-1}a\left(\frac{n}{p}\right) \;. \tag{1.20}$$

(where $a(\alpha) = 0$ if $\alpha$ is not an integer). This means that if $f(z)$ is a simultaneous eigenfunction for all $T(p)$, $(p,N) = 1$, i.e.

$$T(p)f = \lambda_p f, \text{ say,}$$

then

$$a(np) + \psi(p)p^{k-1}a\left(\frac{n}{p}\right) + \lambda_p a(n), \quad \forall (p,N) = 1 \tag{1.21}$$

In particular, if $a(1) = 1 = N$, then

$$a(p) = \lambda_p \tag{1.22}$$

and the Fourier coefficients of $f(z)$ satisfy the simple multiplicative relation

$$a(qp) = a(q)a(p) \tag{1.23}$$

for all primes $p$ and $q$.

The arithmetic interest of relations such as (1.23) (cf. Example B(i)) already makes apparent the power of Hecke's methods.

REMARK 1.8. If $p$ is *not* relatively prime to $N$ the operator $T(p)$ is *not* necessarily normal. Therefore, although the algebra generated by *all* the $T(p)$ on $S_k(N,\psi)$ is still commutative there need *not* exist a basis for $S_k(N,\psi)$ consisting of simultaneous eigenforms for *all* the $T(p)$.

We now introduce the Dirichlet series associated to each $f$ in $S_k(N,\psi)$. If $f$ has the Fourier expansion

$$f(z) = \sum_{n=1}^{\infty} a_n e^{2\pi inz}$$

we set

$$D(s,f) = \sum_{n=1}^{\infty} a_n n^{-s} .$$

This Dirichlet series is (roughly speaking) the Mellin transform of $f$. Indeed, at least formally,

$$\int_0^{\infty} f(iy)y^{s-1}dy = (2\pi)^{-s}\Gamma(s)D(s,f) = L(s,f) . \tag{1.24}$$

If $(r,N) = 1$, and $\chi$ is a primitive character modulo $r$, put

$$g(\chi) = \sum_{x=0}^{r-1} \chi(x)e^{2\pi ix/r} ,$$

$$D(s,f,\chi) = \sum_{n=1}^{\infty} \chi(n)a_n n^{-s} , \tag{1.25}$$

and

(1.26) $$L(s,f,\chi) = (r^2N)^{s/2}(2\pi)^{-s}\Gamma(s)D(s,f,\chi) .$$

Then the principal result of Hecke's theory of Dirichlet series associated with cusp forms is contained in:

THEOREM 1.9.

(i) *Each of the Dirichlet series* $L(s,f,\chi)$ *(associated with* $f$ *in* $S_k(N,\psi)$*) converges in some half-plane, can be analytically continued into the whole plane as an entire function which is bounded in vertical strips* (BV) *and satisfies the functional equation*

(1.27) $$L(s,f,\chi) = i^k\psi(r)\chi(N)g(\chi)^2r^{-1}R(k-s,f|[\sigma]_k,\bar{\chi})$$

*where* $\sigma = \begin{bmatrix} 0 & -1 \\ N & 0 \end{bmatrix}$. *(Note that* $f|[\sigma]_k = f$ *if* $N = 1$.*)*

(ii) $D(s,f)$ *is Eulerian if and only if* $f$ *is an eigenfunction of* $T(p)$ *for all* $p$; *more precisely,* $T(p)f = c_pf$ *for all* $p$ *if and only if*

(1.28) $$D(s,f) = \sum_{n=1}^{\infty} a_n n^{-s} = \prod_p (1 - c_p p^{-s} + \psi(p)p^{k-1-2s})^{-1} .$$

*(We are assuming here that* $f \not\equiv 0$ *and* $a(1) = 1$; *recall also that* $\psi(p) = 0$ *if* $(p,N) \neq 1$.*)*

COROLLARY 1.10. *Suppose* $f_1$ *and* $f_2$ *in* $S_k(N,\psi)$ *are eigenfunctions of* $T(p)$ *for all* $p$ *and suppose* $f_1$ *and* $f_2$ *share the same eigenvalues. Then* $f_1$ *and* $f_2$ *are multiples of one another.*

COROLLARY 1.11 (cf. Example B(i)). *Let* $\Delta(z)$ *again denote the cusp from* $\sum_{n=1}^{\infty} \tau(n)e^{2\pi inz}$ *of weight* 12. *Then:*

(i) $\tau(n)$ *is a multiplicative function (i.e.* $(n,m) = 1$ *implies* $\tau(n)\tau(m) = \tau(nm)$);

(ii) $D(s,\Delta) = \prod_p (1-\tau(p)p^{-s}+p^{11-2s})^{-1}$

(iii) $(2\pi)^{-s}\Gamma(s)D(s,\Delta)$ *is invariant under* $S \to 12-s$.

*Proof.* $S_{12}(SL(2,Z))$ is one dimensional. □

Property (ii) of Corollary 1.11 was conjectured by Ramanujan in 1916 and first proved by Mordell in 1920.

Ramanujan's famous conjecture is that

$$|\tau(p)| \leq 2p^{11/2}$$

for all primes $p$. A generalization of this conjecture (due to Petersson) asserts that if $c_p$ is an eigenvalue of $T(p)$ on $S_k(N,\psi)$ then

$$|c_p| \leq 2p^{(k-1)/2} \tag{1.29}$$

for $(p,N)=1$. Equivalently,

$$\left(1-c_p p^{-s}+p^{k-1-2s}\right) = \left(1-\varepsilon_p p^{\frac{k-1}{2}-s}\right)\left(1-\bar{\varepsilon}_p p^{\frac{k-1}{2}-s}\right)$$

*with* $\varepsilon_p$ *of absolute value* 1 (i.e. the roots of the $1-c_pt+p^{k-1}t^2$ are conjugate). The truth of this conjecture is a consequence of recent work of Deligne ([1] and [2]).

In terms of the Fourier coefficients of $f$ in $S_k(N,\psi)$, (1.29) asserts that

$$a_n = 0\left(n^{\frac{k-1}{2}+\varepsilon}\right).$$

(Cf. the estimate (1.10).)

Soon we shall see that modular forms are special examples of "automorphic forms on $GL(2)$" and that all parts of Hecke's theory, especially Hecke's operators, are more naturally formulated in terms of the representation theory of this group.

## D. Complements to Hecke's Theory

Several natural questions are left *unanswered* by Hecke's work. Two important ones are the following:

(A) Can one characterize the Dirichlet series $L(s,f)$ which arise from forms in $S_k(N,\psi)$? In particular, is there a "converse" to Theorem 1.9(i)?

(B) (Cf. Corollary 1.10.) Suppose $f_1$ and $f_2$ in $S_k(N,\psi)$ are eigenfunctions for $T(p)$ *for all* $(p,N)=1$ and suppose $f_1$ and $f_2$ share the same eigenvalues for these $T(p)$. Are $f_1$ and $f_2$ then multiples of one another?

We deal first with (A). Suppose $\Sigma a_n n^{-s}$ converges in some half-plane, extends holomorphically to the whole s-plane, and satisfies the functional equation

$$L(s) = i^k L(k-s) \tag{1.30}$$

(where $L(s) = (2\pi)^{-s}\Gamma(s)\Sigma a_n n^{-s}$). Then from the identity (1.24) it follows that $L(s)$ is the Mellin transform of some $f$ in $S_k(SL(2,\mathbf{Z}))$, namely $f(z) = \sum_{n=1}^{\infty} a_n e^{2\pi i n z}$, since $\begin{bmatrix} 0 & 1 \\ -1 & 0 \end{bmatrix}$ and $\begin{bmatrix} 1 & 1 \\ 0 & 1 \end{bmatrix}$ generate $SL(2,\mathbf{Z})$.

More generally, Hecke showed that if $R(s) = (2\pi)^{-s}\Gamma(s)\Sigma a_n n^{-s}$ satisfies the functional equation $L(s) = CN^{k/2-s}L(k-s)$, $N>0$, then $\Sigma a_n n^{-s}$ belongs to some modular form of weight $k$ for the discrete subgroup $\Gamma'$ of $SL(2,\mathbf{R})$ generated by $\begin{bmatrix} 1 & 1 \\ 0 & 1 \end{bmatrix}$ and $\begin{bmatrix} 0 & 1 \\ -N & 0 \end{bmatrix}$. *The problem is that* $\Gamma'$ *is rarely a Fuchsian group of the first kind (let alone* $\Gamma_0(N)$), because such groups (in general) have too many generators! (For $N=1$, any form in $S_k(\Gamma_0(N))$ is defined by just *two* "functional equations." The first, for $\begin{bmatrix} 1 & 1 \\ 0 & 1 \end{bmatrix}$ (periodicity), *allows* us to associate to $f$ the Dirichlet series $D(s,f)$. The second, for $\begin{bmatrix} 0 & 1 \\ -1 & 0 \end{bmatrix}$, forces the familiar functional equation (1.30).)

A completely satisfactory response to (A) is due to Weil, whose result is the following:

THEOREM 1.12. *Fix positive integers* $N$ *and* $k$ *and suppose* $D(s) = \sum_{n=1}^{\infty} a_n n^{-s}$ *satisfies the following conditions*:

(i) $D(s)$ *is absolutely convergent for some* $s = k-\delta$, $\delta > 0$;

(ii) *for each primitive character* $\chi$ *modulo* $r$, *with* $(r,N) = 1$, $R^*(s,\chi) = (2\pi)^{-s}\Gamma(s)\Sigma a_n \chi(n) n^{-s}$ *continues to an entire function of* $s$, *bounded in every vertical strip*;

(iii) $L^*(s,\chi)$ *satisfies the functional equation*

$$L^*(s,\chi) = \chi(N)\psi(r)g(\chi)^2 r^{-1}(r^2N)^{k/2-s}L(k-s,\bar{\chi}) .$$

*THEN*: $f(z) = \sum_{n=1}^{\infty} a_n e^{2\pi inz}$ *belongs to* $S_k(N,\psi)$, *i.e.* $\Sigma a_n n^{-s}$ *is the Mellin transform of some* $f$ *in* $S_k(N,\psi)$.

The thrust of this theorem is that the Dirichlet series associated to some form in $S_k(N,\psi)$ can be characterized by simultaneously considering *several related* Dirichlet series. For the obvious reasons we refer to this theorem as "Weil's converse to Hecke theory."

We now return to Question (B).

It is easily seen that the answer (in general) to Question (B) is "no." Indeed, two forms for $\Gamma_0(N)$ may well share eigenvalues for all $T(p)$ with $(p,N) = 1$ and yet be linearly independent.

EXAMPLE 1.13. The two dimensional space $S_{12}(\Gamma_0(2))$ contains $f_2(z) = \Delta(2z)$ as well as $f_1(z) = \Delta(z)$. These two forms have common eigenvalues for all $T(p)$, $p \neq 2$, but are obviously linearly independent.

Example 1.13 suggests that we rephrase Question B as follows:

*Question* (B)$'$. Suppose $f \in S_k(N,\psi)$. What additional assumptions on $f$ are needed to insure that the Fourier coefficients $a_n$, $(n,N) = 1$, completely determine $f$? If $T(p)f = 0$ for all $(p,N) = 1$, when can we conclude that $f$ is identically zero?

This basic question was probably first seriously attacked by Hecke, who proved that *if* $\psi$ *is a primitive character* (i.e. its conductor is precisely $N$) *then* $a_n = 0$ *for* $(n,N) = 1$ *indeed implies* $f$ *is identically zero.* (In this case, every $f$ in $S_k(N,\psi)$ is a "new form" in the sense below.)

Note that Example 1.13 lies outside the scope of this last result of Hecke's. The recent work of Atkin and Lehner, however, deals directly with this example. Since the pathology of this example is precisely what will be of interest to us later on we shall henceforth assume that $\psi = 1$ (mod N).

Clearly Example 1.13 is a special case of the following general phenomenon. Suppose f in $S_k(\Gamma_0(1))$ is an eigenfunction of *every* T(p). Then f also belongs to $S_k(\Gamma_0(N))$, as does $f|\begin{bmatrix} N & 0 \\ 0 & 1 \end{bmatrix}_k$, but

$$\text{(1.31)} \qquad \begin{bmatrix} N & 0 \\ 0 & 1 \end{bmatrix}_k \cdot T(p) = T(p) \cdot \begin{bmatrix} N & 0 \\ 0 & 1 \end{bmatrix}_k$$

if $(p,N) = 1$. Therefore the (linearly independent) forms f(z) and f(Nz) *both* satisfy the equation

$$T(p)f = c(p)f$$

*for all* $(p,N) = 1$. The "explanation" for this is that neither form genuinely "belongs" to $\Gamma_0(N)$. Rather both forms "come from" $\Gamma_0(1)$ and are already "old forms" on $\Gamma_0(N)$.

To make these notions precise, fix m any *proper* divisor of N and d any divisor of $\frac{N}{m}$. Then pick a basis $\{g_j\}$ for $S_k(\Gamma_0(m))$ consisting of eigenfunctions for T(p) with $(p,N) = 1$ and let $S_k^-(\Gamma_0(N)$ denote the subspace of $S_k(\Gamma_0(N))$ spanned by functions of the form $g_j(dz)$. This subspace is preserved by all T(p) with $(p,N) = 1$ because of (1.31). Similarly its orthocomplement in $S_k(\Gamma_0(N))$ (with respect to the Peterson inner product!) is so preserved. This latter subspace of $S_k(\Gamma_0(N))$ we shall denote by $S_k^+(\Gamma_0(N))$. (Observe that this subspace might be the zero space, as in Example 1.13, or the whole space, as in Hecke's case.)

Now pick a basis $\{f_i\}$ for $S_k^+(\Gamma_0(N))$ consisting of eigenfunctions for the T(p) with $(p,N) = 1$ (this is possible since $S_k^+(\Gamma_0(N))$ is preserved by these T(p)). Define a *new form* to be any such $f_i$ and an *old class* to be the class of forms on $S_k(\Gamma_0(N))$ of the form f(dz) with f a fixed *new* form on $\Gamma_0(m)$. (Recall m is any *proper* advisor of N and d any divisor of N/m.) Elements of an old class are called *old forms*.

The significance of new forms is that they are actually eigenfunctions of every $T(p)$. In particular, *for new forms* in $S_k(\Gamma_0(N))$, the response to Question B is "yes"!

The complete story on how the eigenvalues of an eigenform for $\{T(p)\}$, $(p,N) = 1$, do (or do not) determine the form (at least for $f$ in $S_k(\Gamma_0(N))$) is contained in the Theorem 1.14 below. Before stating the theorem, let us agree to call two forms in $S_k(\Gamma_0(N))$ *equivalent* if they have the same eigenvalues for all $T(p)$ with $(p,N) = 1$.

THEOREM 1.14.

(i) *The space* $S_k(\Gamma_0(N)$ *has a basis consisting of new forms and old forms;*

(ii) *Two elements of this basis are equivalent if and only if these elements are old forms and if and only if they belong to the same old class;*

(iii) *Each new form of this basis is an eigenfunction for every* $T(p)$ *with eigenvalue* $0$ *if* $p^2|N$ *and eigenvalue* $\pm p^{k/2-1}$ *if only* $p$ *divides* $N$.

*It follows from (iii) that for the subspace of new forms in* $S_k(\Gamma_0(N))$ *there does exist a basis of forms which are simultaneous eigenfunctions for all the* $T(p)$. *In particular, if* $F(z) = \sum_{n=1}^{\infty} a(n)e^{2\pi inz}$ *is such a basis element, and* $a(1) = 1$, *then*

$$a(np) - a(n)a(p) + p^{k-1}a\left(\frac{n}{p}\right) = 0, \quad \textit{if} \quad (p,N) = 1$$

*and*

$$a(np) - a(n)a(p) = 0, \quad \textit{if} \quad p|N .$$

The full thrust of Theorem 1.14 is best appreciated only after it is reinterpreted (in Section 5) in terms of the representation theory of $GL(2,\mathbb{A})$.

NOTES AND REFERENCES

More complete accounts of Hecke's theory can be found in [Shimura], [Ogg], and [Gunning]. Shimura's treatise also contains an exhaustive bibliography. The compliments to Hecke's theory just described appear in [Weil] and [Atkin-Lehner]. Hecke's original papers are collected in [Hecke]. See also [Serre].

In [Deligne] the Ramanujan-Petersson conjecture is reduced to a famous conjecture of Weil's and in [Deligne 2] this conjecture is proved.

## §2. AUTOMORPHIC FORMS AND THE DECOMPOSITION OF $L^2(\Gamma\backslash SL(2,\mathbb{R}))$

Let G denote the group $SL(2,\mathbb{R})$ and $\Gamma$ the congruence subgroup $\Gamma_0(N)$. One purpose of this section is to describe the correspondence between $S_k(\Gamma)$ and a certain finite-dimensional subspace of $L^2(\Gamma\backslash G)$. This correspondence motivates a quite general notion of "automorphic form" which includes as special cases the non-analytic "wave-forms" of Maass as well as the classical holomorphic forms just discussed.

Another purpose of this section is to explain the role automorphic forms play in decomposing the natural representation of G in $L^2(\Gamma\backslash G)$ and to collect some diverse facts concerning this decomposition.

### A. Automorphic Forms as Functions on SL(2, R)

In G, consider the subgroups

$$A = \left\{\begin{bmatrix} a & 0 \\ 0 & a^{-1}\end{bmatrix} : a > 0\right\}, \quad N = \left\{\begin{bmatrix} 1 & u \\ 0 & 1\end{bmatrix}\right\},$$

and

$$K = \left\{\begin{bmatrix} \cos\theta & -\sin\theta \\ \sin\theta & \cos\theta\end{bmatrix} = r(\theta) : 0 \le \theta < 2\pi\right\}.$$

The subgroup $B = NA$ already acts transitively on $\{\mathrm{Im}(z) > 0\}$ since $\begin{bmatrix} y^{1/2} & xy^{-1/2} \\ 0 & y^{-1/2}\end{bmatrix} i = x+iy$. Thus the upper half-plane is identified with G/K, the stability subgroup of G at i being K.

Since $G = BK = NAK$, each $g \in G$ may be expressed in the form

$$g = \begin{bmatrix} a & b \\ c & d\end{bmatrix} = \begin{bmatrix} y^{1/2} & xy^{-1/2} \\ 0 & y^{-1/2}\end{bmatrix}\begin{bmatrix} \cos\theta & -\sin\theta \\ \sin\theta & \cos\theta\end{bmatrix}. \tag{2.1}$$

Hence, assigning to each g of this form the coordinates $(z = x+iy, \theta)$, a convenient parameterization of G (by $x,y,\theta$), $x \in \mathbb{R}, y > 0$, $\theta \in [0,2\pi]$)

is obtained. Observe that

$$z = g(i), \quad \text{and} \quad \theta = \arg(ci+d) \ , \tag{2.2}$$

if $g = \begin{bmatrix} a & b \\ c & d \end{bmatrix}$.

Now we describe a map from $S_k(\Gamma)$ to a space of functions on $G$. For simplicity we assume $k$ is even. Then for each $f \in S_k(\Gamma)$ we define $\phi_f(g)$ on $G$ by

$$\phi_f(g) = f(g(i))\, j(g,i)^{-k} \ . \tag{2.3}$$

Observe that (by the transitivity of $G$) the function $\phi_f(g)$ is identically zero if and only if $f(z)$ is. The resulting map, $f$ goes to $\phi_f$, is a linear one-to-one map from $S_k(\Gamma)$ to a space of functions $\phi$ on $G$ satisfying the following conditions:

(i) $\phi(\gamma g) = \phi(g)$ for all $\gamma \in \Gamma$;

(ii) $\phi(gr(\theta)) = e^{-ik\theta}\phi(g)$ for all $r(\theta) \in K$;

(iii) $\phi(g)$ is bounded, in particular

$$\int_{\Gamma\backslash G} |\phi(g)|^2\, dg < \infty \ ; \tag{2.4}$$

and

(iv) $\phi(g)$ is *cuspidal*, i.e. for $g \in G$ and $\sigma \in SL(2,\mathbf{Z})$,

$$\int_0^1 \phi\left(\sigma \begin{bmatrix} 1 & xh \\ 0 & 1 \end{bmatrix} g\right) dx = 0 \ .$$

(Here $h$ denotes the "width" of the cusp $\sigma(\infty)$ in the sense of Section 1.)

Condition 2.4(i) follows from the computation

$$\begin{aligned} \phi_f(\gamma g) &= f(\gamma g(i))\, j(\gamma g,i)^{-k} \\ &= j(\gamma,g(i))^k f(g(i))\, j(\gamma,g(i))^{-k}\, j(g,i)^{-k} = \phi_f(g) \ . \end{aligned}$$

(Recall that $j$ is a *factor of automorphy*, i.e. $j(g_1 g_2,z) = j(g_1,g_2 z)\, j(g_2,z)$ for all $g_1, g_2 \in G$.) Similarly (ii) follows from the computation

$$\phi_f(gr(\theta)) = f(gr(\theta)(i))\, j(gr(\theta), i)^{-k}$$
$$= f(g(i))\, j(g,i)^{-k}\, j(r(\theta), i)^{-k} = \phi_f(g)e^{-ik\theta} .$$

(Observe that the restriction of $j(g,i)^{-k}$ to K is the character $e^{-ik\theta}$.)

On the other hand, (iii) follows immediately from Lemma 1.5, and

$$\int \phi_f\left(\sigma \begin{bmatrix} 1 & xh \\ 0 & 1 \end{bmatrix} g\right) dx = \int_0^1 f\left(\sigma \begin{bmatrix} 1 & xh \\ 0 & 1 \end{bmatrix} g(i)\right) j\left(\sigma \begin{bmatrix} 1 & xh \\ 0 & 1 \end{bmatrix} g, i\right)^{-k} dx$$
$$= \int_0^1 f(\sigma(z+hx))\, j(\sigma, z+hx)^{-k} dx$$
$$= \int_0^1 f|[\sigma]_k (hx+z)\, dx .$$

Since this last expression is simply the zeroth Fourier coefficient of f at the cusp $s = \sigma(\infty)$ (cf. (1.8)) Condition 2.4(iv) follows from the fact f is a cusp form in the classical sense.

Perhaps a word is in order concerning the Haar measure used on $\Gamma\backslash G$ in 2.4(iii). Utilizing the parameterization $(x,y,\theta)$ in G we normalize Haar measure on G through the formula

$$\int_G \phi(g)\, dg = \frac{1}{2\pi} \int_0^{2\pi}\int_0^{\infty}\int_{-\infty}^{\infty} \phi(x,y,\theta)\, \frac{dx dy}{y^2}\, d\theta . \tag{2.5}$$

Consequently

$$\int_{\Gamma\backslash G} |\phi_f(g)|^2 dg = \iint_F |f(z)|^2\, y^k \frac{dx dy}{y^2} \tag{2.6}$$

where F denotes a fundamental domain for $\Gamma$ in the upper half-plane. The right side of (2.6) is simply the norm of f(z) with respect to the Petersson inner product.

The question remains: what is the *image* of $S_k(\Gamma)$ in $L^2(\Gamma\backslash G)$ under the map $f \to \phi_f$? To answer this we need to recall the *Laplace operator* for G.

Let $R(g)$ denote the unitary representation of $G$ defined in $L^2(\Gamma\backslash G)$ by the formula

$$R(g)\phi(h) = \phi(hg) \ . \tag{2.7}$$

(This is the so-called right regular representation of $G$ in $L^2(\Gamma\backslash G)$ induced from the trivial representation of $\Gamma$.) Consider the matrices

$$\ell_0 = \begin{bmatrix} 0 & -1 \\ 1 & 0 \end{bmatrix}, \quad \ell_1 = \begin{bmatrix} 0 & 1 \\ 1 & 0 \end{bmatrix},$$

$$\text{and} \quad \ell_2 = \begin{bmatrix} 1 & 0 \\ 0 & -1 \end{bmatrix}$$

which span the Lie algebra of $SL(2,R)$. To each one-parameter group $g_j(t) = \exp(t\ell_j)$ in $G$ there corresponds the one-parameter group of unitary operators $U_j(t) = R(g_j(t))$, hence, by Stone's Theorem, a self-adjoint (unbounded) operator $H_j$ in $L^2(\Gamma\backslash G)$ satisfying $U_j(t) = e^{-itH_j}$.

The *Laplace operator for* $G$ *in* $L^2(\Gamma\backslash G)$ (i.e. the Casimir operator for $R(g)$) is

$$\Delta = -1/4\,(H_0^2 - H_1^2 - H_2^2) \ . \tag{2.8}$$

This operator is defined on smooth functions in $L^2(\Gamma\backslash G)$ and in terms of the coordinates $(x,y,\theta)$ is easily seen to assume the form

$$\Delta = -y^2\left(\frac{\partial^2}{\partial x^2} + \frac{\partial^2}{\partial y^2}\right) - y\,\frac{\partial^2}{\partial x \partial \theta} \ ; \tag{2.9}$$

Moreover, (the minimal closed extension of) $\Delta$ is self-adjoint in $L^2(\Gamma\backslash G)$ *and commutes with* $R(g)$. Consequently it follows from Schur's Lemma that the restriction of $\Delta$ to any G-invariant *irreducible* subspace of $L^2(\Gamma\backslash G)$ is a scalar.

We can now characterize the image of $S_k(\Gamma)$ in $L^2(\Gamma\backslash G)$. Let $A_k^2(\Gamma)$ denote the space of functions $\phi$ on $G$ satisfying the following conditions:

$$(2.10)\quad \begin{aligned}&\text{(a)}\ \phi(\gamma g) = \phi(g)\ \text{ for all }\ \gamma \in \Gamma;\\ &\text{(b)}\ \phi(gr(\theta)) = \phi(g)e^{-ik\theta}\ \text{ for all }\ r(\theta)\in K;\\ &\text{(c)}\ \Delta\phi = -\frac{k}{2}\left(\frac{k}{2}-1\right)\phi;\ \text{ and}\\ &\text{(d)}\ \phi\ \text{ is bounded and cuspidal.}\end{aligned}$$

Note that $A_k^2(\Gamma)$ is contained in $L^2(\Gamma\backslash G)$ by 2.10(d).

PROPOSITION 2.1. *The formula*

$$\phi_f(g) = f(g(i))\,j(g,i)^{-k}$$

*describes an isomorphism between* $S_k(\Gamma)$ *and* $A_k^2(\Gamma)$.

*Proof.* We have only to show that $\phi_f$ satisfies 2.10(c) and that the map is onto (cf. 2.4). So first suppose $f \in S_k(\Gamma)$. Since $\phi_f(g) = y^{k/2}f(z)e^{-ik\theta} = \phi_f(z,\theta)$, a straight forward computation shows that

$$\begin{aligned}\Delta\phi_f(g) &= \left(-y^2\left(\frac{\partial^2}{\partial x^2}+\frac{\partial^2}{\partial y^2}\right) - y\frac{\partial^2}{\partial x\partial\theta}\right)\phi(z,\theta)\\ &= -e^{-ik\theta}y^{(k/2)+2}\left(\frac{\partial^2}{\partial x^2}+\frac{\partial^2}{\partial y^2}\right)f(z)\\ &\quad + (ik)e^{-ik\theta}y^{k/2+1}\left(\frac{\partial}{\partial x}+i\frac{\partial}{\partial y}\right)f(z)\\ &\quad -\frac{k}{2}\left(\frac{k}{2}-1\right)y^{k/2}f(z)e^{-ik\theta}\,.\end{aligned}\tag{2.11}$$

But $f(z)$ is holomorphic by assumption. Therefore the first two terms on the right side of (2.11) vanish and 2.10(c) is established.

Now suppose $\phi(g)$ is arbitrary in $A_k^2(\Gamma)$ and put

$$f(z) = \phi(g)\,j(g,i)^k\,.\tag{2.12}$$

(Here $g$ is any element at all of $G$ such that $g(i) = z$.) From 2.10(b) it follows that $f(z)$ is well-defined. On the other hand $\phi_f$ certainly equals $\phi$. Therefore to prove this proposition it remains only to check that

$f_\phi(z) \in S_k(\Gamma)$. Since this verification requires facts not yet introduced we shall simply sketch the argument for the sake of completeness.

Let $L_0^2(\Gamma\backslash G)$ denote the space of $\phi$ in $L^2(\Gamma\backslash G)$ satisfying 2.4(iv) for almost every $g \in G$. By hypothesis, $\phi \in L_0^2(\Gamma\backslash G)$. On the other hand (as we shall see in Theorem 2.6(ii)),

$$L_0^2(\Gamma \quad G) = \bigoplus_j H^j$$

where each $H^j$ is invariant for the right action of G *and* irreducible. Consequently

$$\phi = \sum \phi_\ell$$

where $\phi_\ell \in H^\ell$.

Note now that each $\phi_j$ satisfies 2.10(b) since $\phi$ does. Recalling that $\Delta$ operates as a scalar on each irreducible subspace of $L^2(\Gamma\backslash G)$, we also have

$$\Delta\phi_\ell = -\frac{k}{2}\left(\frac{k}{2}-1\right)\phi_\ell$$

for each $\ell$, since

$$\lambda_\ell(\phi,\phi_\ell) = (\phi,\Delta\phi_\ell) = (\Delta\phi,\phi_\ell) = -\frac{k}{2}\left(\frac{k}{2}-1\right)(\phi,\phi_\ell)$$

(by hypothesis and the self-adjointness of $\Delta$). From this it may be deduced that $\phi$ is a linear combination of "lowest weight vectors" from irreducible subspaces of $L_0^2(\Gamma\backslash G)$ equivalent to the discrete series representation $\pi_k^+$ (see Section B). Moreover, each such vector (and hence $\phi$ itself) can be shown to be annihilated by a certain first order differential operator in $(x,y,\theta)$ which in turn implies that $f_\phi(z)$ is annihilated by the first order operator

$$\frac{\partial}{\partial x} + i\frac{\partial}{\partial y}\ .$$

Consequently $f_\phi(z)$ is holomorphic. (The details of this argument appear in Section 4 of Chapter 1 of [Gelfand-Graev-Pyatetskii Shapiro].)

That $f_\phi(z)$ is automorphic of weight $k$ and a cusp form follows immediately by reversing the computations used earlier to show that $\phi_f \in A_k^2(\Gamma)$ whenever $f \in S_k(\Gamma)$. □

According to Proposition 2.1, classical (holomorphic) cusp forms are examples of right K-finite eigenfunctions of $\Delta$ in $L^2(\Gamma\backslash G)$. (*Right K-finite* means that the space of functions on G spanned by the right translates of $\phi(g)$ by k in K is finite-dimensional.) Now from the point of view of the spectral decomposition of $\Delta$ there is nothing particularly special about the eigenvalue $-k/2(k/2-1)$ (cf. 2.10(c)). Therefore it is natural to make the following definition.

DEFINITION 2.2. *A $\Gamma$-automorphic form* $\phi$ on G is a (smooth) function on G satisfying the following properties:

(2.13)
(a) $\phi(\gamma g) = \phi(g)$ for all $\gamma \in \Gamma$;
(b) $\phi$ is right K-finite;
(c) $\phi$ is an eigenfunction of $\Delta$; and
(d) $\phi$ satisfies a certain growth condition:

there are constants C and N such that

$$|\phi(z,\theta)| \leq Cy^N \ (y \to +\infty) .$$

(We say that $\phi$ is *slowly increasing*.)

If, in addition to satisfying properties 2.13(a)-(d), $\phi$ is cuspidal in the sense of 2.4(iv), then $\phi$ is called a $\Gamma$-*cusp form*. Of course elements of $S_k(\Gamma)$ (or rather $A_k^2(\Gamma)$) constitute the fundamental examples of $\Gamma$-automorphic forms on G. Others include:

EXAMPLE 2.3 (*Non-holomorphic cusp forms*). Let $W_s(\Gamma)$ denote the space of smooth right K-*invariant* functions $\phi$ on G which are bounded and such that

$$\Delta\phi = \frac{1-s^2}{4}\phi .$$

Then each $\phi$ in $W_s(\Gamma)$ is obviously an automorphic form on G in the sense of Definition 2.2. Furthermore, the formula

$$f_\phi(z) = \phi(g(i)) ,$$

where g in G maps i to z, establishes an isomorphism between $W_s(\Gamma)$ and the finite-dimensional space of complex-valued functions f(z) on $\{\mathrm{Im}(z) > 0\}$ satisfying the following properties:

(i) $f(\gamma z) = f(z)$ for all $\gamma \in \Gamma$ (i.e. f(z) is an automorphic *function*!);

(2.14) (ii) $\Delta^* f = \frac{1-s^2}{4} f$, where $\Delta^* = -y^2\left(\frac{\partial^2}{\partial x^2} + \frac{\partial^2}{\partial y^2}\right)$ is the Laplace-Beltrami operator for the upper half-plane (f is *real* analytic, not holomorphic);

(iii) f(z) is bounded.

These functions were first systematically studied by H. Maass and called by him "wave forms."

REMARK 2.4. In Condition (ii) of Example 2.3 *we must assume* s *to be a pure imaginary number or a non-zero real number between* $-1$ *and* 1. This is because the Laplace-Beltrami operator is non-negative.

Soon we shall see that $W_s(\Gamma)$ is non-trivial for infinitely many values of s. However, to the best of our knowledge, almost nothing is known concerning those specific values of s for which $W_s(\Gamma)$ is non-trivial. (For certain *proper* subgroups of SL(2,Z), a *few* examples are known; cf. Theorem 2.14 below.)

EXAMPLE 2.5 (*Real Analytic Eisenstein Series*). Suppose $\Gamma = SL(2,\mathbf{Z})$, $\mu \in \mathbf{C}$, and $z = x+iy, y > 0$. If $f_\mu(z) = y^\mu$ the series

$$(2.15) \qquad E^*(z,\mu) = \sum_{\gamma \in \Gamma \cap N\backslash\Gamma} f_\mu(\gamma z) = \sum_{\substack{(c,d)=1 \\ c,d \in \mathbf{Z}}} \frac{y^\mu}{|cz+d|^{2\mu}}$$

is known to converge for $\operatorname{Re}(\mu) > 1$ and to have a meromorphic continuation to the whole $\mu$-plane with a simple pole at $\mu = 1$. The resulting function of $z$ is known as the *Eisenstein series* of index $\mu$ attached to the cusp $\infty$ of $\Gamma$ and its significance is as follows. If $\mu = 1/2 + s/2$, and $s$ is pure imaginary, then the formula

$$E(g,s) = E^*\left(g(i), \frac{1}{2} + \frac{s}{2}\right) \tag{2.16}$$

describes a right K-invariant $\Gamma$-automorphic form on G which is *not* a cusp form. Its eigenvalue for $\Delta$ is $\frac{1-s^2}{4}$.

Generalizations of $E(g,s)$ are discussed in Section 8 and play a fundamental role in the theory of automorphic forms. In fact Examples 2.3 and 2.5, together with $A_k^2(\Gamma)$, essentially exhaust all possible automorphic forms on G.

## B. Automorphic Forms and the Decomposition of $L^2(\Gamma\backslash G)$

As before, let $R(g)$ denote the right regular representation of G in $L^2(\Gamma\backslash G)$.

From Proposition 2.1 we know that there is a one-to-one correspondence between the elements of $S_k(\Gamma)$ and certain eigenfunctions of $\Delta$ in $L^2(\Gamma\backslash G)$. Similarly, by Example 2.3, there is a one-to-one correspondence between certain other eigenfunctions of $\Delta$ and real-analytic waveforms.

On the other hand, the restriction of $\Delta$ to irreducible subrepresentations of $R(g)$ is known *a priori* to operate as a scalar. Therefore one must expect automorphic forms to play a role in the decomposition of $R(g)$. That they do provides a specific instance of the general principle whereby eigenspaces of an operator commuting with a given representation reduce that representation. In the context of $R(g)$ this operator is clearly $\Delta$. And since the restriction of $R(g)$ to K is completely reducible (K being compact) there is certainly no loss of generality in requiring these eigenfunctions to be right K-finite. Therefore *the problem of decomposing* $R(g)$

*is virtually indistinguishable from the problem of constructing automorphic forms.*

In this subsection we shall directly consider the representation theoretic problem of decomposing $R(g)$.

We start by describing the irreducible unitary representations of $SL(2,R)$.

(a) *The class* 1 *principal series* $\pi_s^+(g)$.

These representations are induced from the unitary characters

$$\begin{bmatrix} a & u \\ 0 & a^{-1} \end{bmatrix} \mapsto |a|^s \ (\mathrm{Re}\,(s) = 0) \tag{2.17}$$

of $B$. Hence they are described as right translation operators in the Hilbert space $H^+(s)$ consisting of measurable functions $\phi$ on $G$ such that

$$\phi\left(\begin{bmatrix} a & u \\ 0 & a^{-1} \end{bmatrix} g\right) = |a|^{s+1}\,\phi(g) \tag{2.18}$$

and

$$\int_K |\phi(k)|^2\,dk = \|\phi\|^2 < \infty\ . \tag{2.19}$$

We recall that $\pi_s^+$ and $\pi_{-s}^+$ are unitarily equivalent.

(b) *The non-class* 1 *principal series* $\pi_s^-(g)$.

These representations are induced from the characters

$$\begin{bmatrix} a & u \\ 0 & a^{-1} \end{bmatrix} \mapsto \mathrm{sgn}\,(a)\,|a|^s,\ \ \mathrm{Re}\,(s) = 0\ ,$$

and are irreducible if and only if $s \neq 0$. Hence we assume $s \neq 0$. Again, $\pi_s^-$ and $\pi_{-s}^-$ are equivalent.

(c) *The complementary series* $\pi_s^c(g)$.

These unitary representations result from induction from the *non*-unitary characters

$$\begin{bmatrix} a & u \\ 0 & a^{-1} \end{bmatrix} \to |a|^s$$

of $B$ *with* $s$ *a non-zero real number between* $-1$ *and* $1$. Of course the inner product in $H(s)$ invariant for $\pi_s^c(g)$ is longer given by (2.19) but by a more complicated formula which we omit. Suffice it to say that such an invariant inner product *does* exist and results from the positive-definiteness of the analytic continuation of the operator intertwining the equivalent representations $\pi_s^+$ and $\pi_{-s}^+$.

(d) *The discrete series* $\pi_k^{\pm}(g)$.

These representations are subrepresentations of certain non-unitary representations again induced from B but may be conveniently described as follows. For each integer $k > 1$, let $H(k)$ denote the Hilbert space of holomorphic functions in the upper half-plane satisfying

$$\iint_{\mathrm{Im}(z)>0} |f(x+iy)|^2\, y^k \frac{dxdy}{y^2} < \infty\,. \tag{2.20}$$

Then

$$\pi_k^+(g)f(z) = (bz+d)^{-k} f\left(\frac{az+c}{bz+d}\right) \tag{2.21}$$

if $g = \begin{bmatrix} a & b \\ c & d \end{bmatrix}$. The representations $\pi_k^-(g)$ are defined similarly by replacing the upper half-plane by the lower and $y^k$ by $|y|^k$.

(e) *The trivial representation*.

It is a well-known fact due to Bargmann that the representations just described exhaust the totality of irreducible unitary representations of G. Therefore each irreducible subrepresentation of $R(g)$ is equivalent to some one of these and the question remains which irreducible representations so appear.

Observe that since $\Gamma\backslash G$ is non-compact there is no reason to expect the decomposition of $R(g)$ to be entirely discrete. Moreover it should not seem surprising that this discrete spectrum is (essentially) exhausted by the cuspidal functions in $L^2(\Gamma\backslash G)$.

To make matters precise, suppose $\Gamma$ is $SL(2,\mathbb{Z})$ and let $L_0^2(\Gamma\backslash G)$ denote the subspace of $\phi$ in $L^2(\Gamma\backslash G)$ satisfying the cuspidal condition 2.4(iv). Note that for $SL(2,\mathbb{Z})$ this condition is

$$\int_{N\cup\Gamma\backslash N} \phi(ng)\,dn = 0 \quad \text{for} \quad \text{a.e.g.}$$

Since this condition is obviously invariant under the right action of G the space of *cusp forms* $L_0^2(\Gamma\backslash G)$ is invariant for $R(g)$.

THEOREM 2.6. *Let* $L_d^2(\Gamma\backslash G)$ *denote the direct sum of all subspaces of* $L^2(\Gamma\backslash G)$ *irreducibly invariant under the action of* G. *Then*:

(i) $L_d^2(\Gamma\backslash G)$ *is the direct sum of* $L_0^2(\Gamma\backslash G)$ *and the one-dimensional subspace of constant functions*;

(ii) *The spectrum of* $L_0^2(\Gamma\backslash G)$ *has finite multiplicity, i.e. in the restriction of* $R(g)$ *to* $L_0^2(\Gamma\backslash G)$ *each irreducible representation of* G *occurs at most finitely many times*;

(iii) *The restriction of* $R(g)$ *to the orthocomplement of* $L_d^2(\Gamma\backslash G)$ *is the continuous sum of the principal series representations* $\pi_s^+(g)$, *i.e. if* $R_c(g)$ *denotes the restriction of* $R(g)$ *to the orthocomplement of* $L_d^2(\Gamma\backslash G)$ *then*

$$R_c(g) = \bigoplus \int_0^\infty \pi_s^+(g)\,ds \ . \tag{2.22}$$

This theorem appears as a corollary to the discussions of Sections 8 and 9, and, as already remarked, is equivalent to the spectral decomposition of $\Delta$. It's proof involves showing that any slowly increasing function $f(z)$ on $\{\mathrm{Im}(z) > 0\}$ which is orthogonal to all cusp forms may be expressed as an integral average

$$\int_{\substack{\mathrm{Re}(s)=0\\ \mathrm{Im}(s)>0}} \hat{f}\left(\frac{1}{2}+\frac{s}{2}\right) E(z,s)\,ds$$

of real analytic Eisenstein series $E(z,s)$.

REMARK 2.7. Suppose $\Gamma$ is Hecke's subgroup $\Gamma_0(N)$ and m denotes the number of inequivalent cusps of $\Gamma$. Then the continuous sum (2.22) becomes an *m-dimensional* continuous sum and Parts (i) and (ii) of Theorem 2.6 remain unchanged. Now suppose $\Gamma$ is an arbitrary Fuchsian group of the first kind. Then Part (i) is no longer valid since in this generality infinite dimensional representations of G may occur in $L_d^2(\Gamma\backslash G)$ *outside* the space of cusp forms. Indeed square-integrable $\Gamma$-automorphic forms which are not cusp forms arise as residues at the poles of Eisenstein series (cf. Section 8) and for general $\Gamma$ these poles may be more interesting than the pole at 1 is for $\Gamma = \Gamma_0(N)$.

REMARK 2.8. Although Theorem 2.6 completely describes the *continuous* spectrum of R(g) it tells us nothing about the *discrete* spectrum except that this spectrum has finite multiplicities. The question remains which representations of G occur discretely and with what multiplicity? The only general result along these lines is a "duality theorem" essentially due to Gelfand, Fomin, and Pyatetskii-Shapiro. To describe it we need first to collect several basic facts from the representation theory of G. (For more complete treatments of these well-known facts the reader is referred to the "Notes and References" at the end of this section.)

LEMMA 2.9. *If* $H^{\pm}(s)$ (resp $H^c(s)$) *denotes an irreducible subspace of* $L^2(\Gamma\backslash G)$ *equivalent to* $\pi_s^{\pm}(g)$ (resp. $\pi_s^c(g)$) *then*

$$\Delta\phi = \frac{1-s^2}{4}\,\phi \tag{2.23}$$

*on* $H^{\pm}(s)$ (resp. $H^c(s)$). *Similarly*

$$\Delta\phi = -\frac{k}{2}\left(\frac{k}{2}-1\right)\phi \tag{2.24}$$

*on* $H^{\pm}(k)$.

*Proof* (Sketch). This lemma is equivalent to the statement that the Casimir operator for the irreducible representation $\pi_s^{\pm}, \pi_s^c$, or $\pi_k^{\pm}$ (cf. (2.8)) is described by (2.22) or (2.23). But this last fact follows from a straightforward computation using the formulas

$$H_j = \lim_{t\to 0} \frac{\pi(\exp t\ell_j) - I}{t} = \frac{d}{dt}(\pi(\exp t\ell_j))\Big|_{t=0}$$

$$\Delta = -1/4\,(H_0^2 - H_1^2 - H_2^2)$$

and convenient realizations of $\pi_s^{\pm}$ and $\pi_s^{c}$. □

To describe the result of Gelfand, Fomin, Pyatetskii-Shapiro we need to recall the following basic facts from the representation theory of G:

(i) The restriction to K of an irreducible unitary representation $\pi$ of G contains a given irreducible representation of K at most once;

(ii) This restriction contains the *trivial* representation of K if and only if $\pi$ is some $\pi_s^+$ or $\pi_s^c$: These representations have precisely one K-fixed vector in their space and comprise the so-called *class* 1 *representations* of G;

(iii) *The restriction of* $\pi_k^{\pm}$ to K contains the K-representation $r(\theta) \to e^{\mp ik\theta}$.

We can now state and prove:

THEOREM 2.10. *The representation* $\pi_k^{\pm}$ *of* G (resp. $\pi_s^+$ *or* $\pi_s^c$) *occurs in* $R_0(g)$ (*the restriction of* R(g) *to* $L_0^2(\Gamma\backslash G)$) *with multiplicity equal to the dimension of* $S_k(\Gamma)$ (resp. $W_s(\Gamma)$).

*Proof.* By Theorem 2.6 we can write

$$R_0(g) = \bigoplus_j \pi^j \tag{2.25}$$

where $\pi^j$ is irreducible and acts on some subspace $H^j$ of $L^2(\Gamma\backslash G)$ by right translation. So consider the representation of K defined in $H^j$ by restriction to K, i.e. by the operators $\pi^j(r(\theta))$, $r(\theta) \in K$. Since $K = SO(2)$ is compact and abelian,

$$H^j = \bigoplus \{\phi_\ell^j(g)\}\ ,$$

where each $\phi_\ell^j$ is in the domain of $\Delta$ and such that $\phi_\ell^j(gr(\theta)) = e^{-i\ell\theta}\phi_\ell^j(g)$. Now suppose $H^j$ is equivalent to $\pi_k^+$, say. Then exactly one $\phi_\ell^j$ in $H^j$, namely $\phi_k^j$ (the "lowest weight vector for $\pi_k^+$"), simultaneously satisfies the conditions

$$\phi(g(r(\theta)) = e^{-ik\theta}\phi(g) \tag{2.26}$$

and

$$\Delta\phi = -\frac{k}{2}\left(\frac{k}{2}-1\right)\phi \ . \tag{2.27}$$

Consequently, by Proposition 2.1, exactly one $\phi_\ell^j$ in $H^j$ "belongs" to $S_k(\Gamma)$. Hence $m_k^+$, the multiplicity of $\pi_k^+$ in $R_0(g)$, is *at most* the dimension of $S_k(\Gamma)$.

To obtain the reverse inequality we let f denote any non-zero element of $S_k(\Gamma)$. Then $\phi_f(g) = f(g(i))\, j(g,i)^{-k}$ belongs to $L_0^2(\Gamma\backslash G)$ and satisfies (2.26) and (2.27), again by Proposition 2.1.

By the complete reducibility of $R_0(g)$ (cf. 2.24),

$$\phi_f = \sum_j \phi_j \ ,$$

with each $\phi_j$ a smooth function in some irreducible subspace $H^j$ of $L_0^2(\Gamma\backslash G)$. These $\phi_j$ again satisfy (2.26), and $\Delta\phi_j = \lambda_j\phi_j$ for *some* $\lambda_j$ by the irreducibility of $H^j$. Actually, by the self-adjointness of $\Delta$, $(\phi,\Delta\phi_j) = (\Delta\phi,\phi_j)$, so $\lambda_j$ must equal $-k/2(k/2-1)$. Consequently $H^j$ must be equivalent to $\pi_k^+$. Since this implies that the dimension of $S_k(\Gamma)$ is at most $m_k^+$, the proof of the theorem is complete for $\pi_k^+$.

The proof for $\pi_s^\pm$ and $\pi_s^c$ is similar except that in these cases one deals with right K-*invariant* $\phi_j$ and appeals to Example 2.3. □

Observe that if $\begin{bmatrix}-1 & 0\\ 0 & -1\end{bmatrix} \in \Gamma$ then $R_0(g)$ can never contain the representations $\pi_s^c$ or $\pi_k^\pm$ with k odd.

## C. Miscellaneous Results Concerning the Decomposition of $L^2(\Gamma\backslash SL(2,\mathbf{R}))$

As far as principal and complementary series are concerned the duality theorem of the last subsection yields very little explicit information. This is because for any fixed value of $s$ very little is known concerning the dimension of $W_s(\Gamma)$.

The following theorem shows that we at least know $W_s(\Gamma)$ is non-empty for infinitely many (mysterious) values of $s$.

THEOREM 2.11. *In the decomposition of* $R_0(g)$, *infinitely many class* 1 *representations occur. Equivalently,* $W_s(\Gamma)$ *is non-empty for infinitely many permissible values of* $s$.

*Proof* (Sketch). Letting $R_0(g)$ denote right translation in the space of cusp forms $L_0^2(\Gamma\backslash G)$ we again write

$$R_0(g) = \bigoplus_j \pi^j \tag{2.28}$$

where each $\pi^i$ is an irreducible unitary representation of $G$. Clearly the subspace of K-*fixed* vectors in $L_0^2(\Gamma\backslash G)$ coincides with the Hilbert space $L_0^2(\Gamma\backslash G/K) = L_0^2(\Gamma\backslash\{\mathrm{Im}(z) > 0\})$. However, it can be shown that this latter space is infinite-dimensional. Therefore, since each $\pi^j$ in (2.28) contains the trivial representation of $K$ at most once, infinitely many of these $\pi^j$ must be class $1$. □

This theorem, of course, does not tell us whether representations of *both* the principal *and* complementary series occur. More generally, it certainly does not provide specific values of $s$ for which $W_s(\Gamma)$ is non-empty.

CONJECTURE 2.12. If $\pi_s^+$ belongs to the principal series of representations, then $\pi_s^+$ occurs in $L^2(SL(2,\mathbf{Z})\backslash G)$ *at most once.*

CONJECTURE 2.13. No representation at all of the complementary series for $G$ occurs in $L^2(SL(2,Z)\backslash G)$.

Interesting experimental evidence supporting the truth of these conjectures has recently been obtained by Cartier. The second conjecture is actually a theorem and is attributed to Selberg (see [Roelcke 1]).

The following special result is a sample of possible consequences of Jacquet-Langland's theory and its proof is postponed until Section 7.

THEOREM 2.14. *The principal series representation* $\pi_s^+$, *with*

$$s = \frac{2\pi i}{\log(\sqrt{2}-1)},$$

*occurs discretely in* $L^2(\Gamma_0(2)\backslash G)$.

A last miscellaneous result concerns the discrete spectrum of an arbitrary Fuchsian group of the first kind and was communicated to us by R. Hotta.

Recall that in the generality of an arbitrary Fuchsian group non-trivial representations of $G$ can occur *discretely* outside the space of cusp forms. The thrust of the result below is that these representations never belong to the discrete series for $G$. This fact appears naturally as a special consequence of Harish-Chandra's general theory of discrete series and cusp forms for arbitrary semi-simple groups. Thus we state it without proof rather than digress from the subject matter of this section.

THEOREM 2.15. *Suppose* $\Gamma$ *is an arbitrary Fuchsian group of the first kind and* $\pi_k$ *is a representation of the discrete series with* $k > 2$. *Then if* $\pi_k$ *occurs in* $L_d^2(\Gamma\backslash G)$ *it occurs in* $L_0^2(\Gamma\backslash G)$.

CONCLUDING REMARK. Theorem 2.15 also follows from the theory of Eisenstein series. Indeed this theory implies that each "non-cuspidal" subrepresentation of $L_d^2(\Gamma\backslash G)$ "imbeds" in some $\pi_s^c(g)$ with $s \in [0,1]$. (Cf. Sections 4.A and 8.C.)

## NOTES AND REFERENCES

We recommend [Gelfand-Graev-Pyatetskii-Shapiro] for further discussion of the basic notions of this section (at least for the case of compact quotient). The connection between automorphic forms and group representations seems first to have been made explicit by [Gelfand-Fomin] in 1952. Definition 2.2 is a special case of the general notion of an automorphic form on an arbitrary semi-simple group introduced in [Harish-Chandra].

The theory of real analytic Eisenstein series was introduced in [Selberg] for Fuchsian groups and subsequently significantly generalized in [Langlands]. A detailed treatment of Selberg's theory (in particular, Theorem 2.6) has recently been given in [Kubota]. Further discussion of Eisenstein series and the continuous spectrum is to be found in Section 8.

The experimental evidence supporting Conjectures 2.12 and 2.13 appears in [Cartier].

## §3. AUTOMORPHIC FORMS AS FUNCTIONS ON THE ADELE GROUP OF GL(2)

In Section 5 we shall realize classical cusp forms not as functions on $SL(2,R)$ but as irreducible representations of the adele group of $GL(2)$ over $Q$. To make this correspondence explicit we shall (in Section 4) collect several facts from the representation theory of $GL(2)$ over local and global fields. This background is necessary if one wants to understand the precise sense in which the correspondence between cusp forms and representations fails to be one-to-one. It also provides an appropriate setting for the work of Jacquet and Langlands where the classical theory of automorphic forms is attacked from the point of view of group representations.

Our goal in this section is to describe how classical cusp forms correspond first to special *functions* on the adele group of $GL(2)$.

### A. Basic Notions

Throughout this subsection, $p$ will denote an arbitrary place of $Q$. By $Q_p$ we shall denote the completion of $Q$ at $p$ and by $G_p$ the local group $GL(2,Q_p)$. If $p$ is infinite then $G_p$ is $GL(2,R)$. If $p$ is finite we let $O_p$ denote the ring of integers of $Q_p$ and set $K_p$ equal to $GL(2,O_p)$. Each $K_p$ is then a maximal compact *open* subgroup of $G_p$. (If $p = \infty$ we take $K_p$ to be $O(2,R)$.)

The adele ring of $Q$ will be denoted $A$ (or $A(Q)$ if this reference to the global field is necessary). The group

$$G_A = GL(2,A)$$

is then the direct product of the $G_p$ above restricted with respect to $K_p$.

Its center is

$$Z_A = \left\{ \begin{bmatrix} t & 0 \\ 0 & t \end{bmatrix} : t \in A^x \right\}.$$

By $G_Q$ we denote the discrete subgroup $GL(2,Q)$ of $G_A$. Since $G_A$ is unimodular, the quotient space $G_Q \backslash G_A$ inherits a natural $G_A$-invariant measure. The center of $G_Q$ is $Z_Q = \left\{ \begin{bmatrix} a & 0 \\ 0 & a \end{bmatrix} : a \in Q^x \right\}$ so $Z_Q \backslash Z_A$ is naturally isomorphic to the *idele class group* $Q^x \backslash A^x$. The measure of the homogeneous space $Z_A G_Q \backslash G_A$ is finite.

We recall now the principle of strong approximation for $SL(2)$ over $Q$. This principle implies (see "Notes and References") that

(3.1) $$G_A = G_Q G_\infty^+ K'_0$$

if $G_\infty^+ = GL^+(2,R)$, $K'_0 = \prod_{p<\infty} K'_p$, and $K'_p$ is any choice of open subgroup of $K_p$ such that $K'_p = K_p$ for almost every $p$ *and* such that the determinant map from $K_p$ to $O_p^x$ is surjective *for every* $p$. In particular, (3.1) will hold when

(3.2) $$K'_p = K_p^N = \left\{ \begin{bmatrix} a & b \\ c & d \end{bmatrix} \in K_p : c \equiv 0(N) \right\},$$

$N$ any positive integer. In this case we write $K_0^N$ for $K'_0$.

The analogue of (3.1) for $GL(1)$ is

(3.3) $$A^x = Q^x R_+^x \prod_{p<\infty} Q_p .$$

This expression is equivalent to the fact that $Z$ has class number 1.

By a *grossencharacter of* $Q$ we shall understand a unitary character of the idele class group of $Q$, i.e. a character of $A^x$ trivial on $Q^x$. Observe that by (3.3) each character of $(Z/NZ)^x$ canonically determines a grossencharacter of $Q$. Indeed suppose $\psi$ is such a character. Then $\psi$ determines a character $\psi_p$ of $O_p^x$ by composition with the natural homomorphism from $O_p^x$ to $(Z/NZ)^x$. The product $\prod_{p<\infty} \psi_p$ then describes a character of $\prod_{p<\infty} O_p^x$, hence a character of $A^x$ (trivial on $Q^x R_+^x$!) by (3.3).

Now fix $f \in S_k(N,\psi)$. Using (3.1), and the fact that

$$\begin{bmatrix} a & b \\ c & a \end{bmatrix} \mapsto \psi_p(a)$$

determines a character of $K_p^N$, we can define a function $\phi_f(g)$ on $G_A$ by

$$\phi_f(g) = f(g_\infty(i))\, j(g_\infty, i)^{-k} \psi(k_0) \tag{3.4}$$

if $g = \gamma g_\infty k_0$ and $\Pi \psi_p = \psi$. This function is well defined since

$$G_Q \cap G_\infty^+ \prod K_p^N = \Gamma_0(N) \ . \tag{3.5}$$

Indeed f is assumed to belong to $S_k(N,\psi)$ and consequently $\phi_f(\gamma g) = \phi_f(g)$ for all $\gamma \in \Gamma_0(N)$.

Observe that if $\psi = 1$, and we regard $\phi_f$ as a function of SL(2,R) alone, then $\phi_f$ coincides with the $\phi_f$ of Proposition 2.1. In general, using strong approximation, it is not difficult to establish the following proposition.

PROPOSITION 3.1. *The map* f(z) *goes to* $\phi_f(g)$ *describes an isomorphism between* $S_k(N,\psi)$ *and the space of functions* $\phi$ *on* $G_A$ *satisfying the following conditions*:

(i) $\phi(\gamma g) = \phi(g)$ *for all* $\gamma \in G_Q$;

(ii) $\phi(gk_0) = \phi(g)\psi(k_0)$ *for all* $k_0 \in \Pi K_p^N$;

(iii) $\phi(gr(\theta)) = e^{-ik\theta}\phi(g)$ *if* $r(\theta) = \begin{bmatrix} \cos\theta & -\sin\theta \\ \sin\theta & \cos\theta \end{bmatrix}$;

(iv) *the function* $\phi$, *viewed as a function of* $G_\infty^+$ *alone, satisfies the differential equation*

$$\Delta\phi = -\frac{k}{2}\left(\frac{k}{2}-1\right)\phi \ ; \tag{3.6}$$

(v) $\phi(zg) = \phi(gz) = \psi(z)\phi(g)$ *for all* $z \in Z_A$;

(vi) $\phi$ *is slowly increasing, i.e. for every* $c > 0$, *and compact subset* $\omega$ *of* $G_A$, *there exist constants* C *and* N *such that*

$$\phi\left(\begin{bmatrix} a & 0 \\ 0 & 1 \end{bmatrix} g\right) \leq C|a|^N$$

*for all* $g \in \omega$, *and* $a \in A^x$ *with* $|a| > c$: *This is the natural adelic version of the classical notion described earlier in the context of Definition 2.2;*

(vii) $\phi$ *is cuspidal, i.e.*

$$\int_{Q\backslash A} \phi\left(\begin{bmatrix} 1 & x \\ 0 & 1 \end{bmatrix} g\right) dx = 0 \text{ for almost every } g.$$

We remark that the proof of Proposition 3.1 involves showing that the "cuspidal" (resp. slowly increasing) condition for $\phi_f$ implies that $f(z)$ vanishes (resp. is regular) at every cusp of $\Gamma$.

DEFINITION 3.2. Let $L^2(G_Q\backslash G_A,\psi)$ denote the Hilbert space of measurable functions $\phi$ on $G_A$ such that

(i) $\phi(\gamma g) = \phi(g)$ for all $\gamma \in G_Q$;

(3.7) (ii) $\phi(gz) = \phi(g)\psi(z)$ for all $z \in Z_A$; and

(iii) $\int_{Z_A G_Q\backslash G_A} |\phi(g)|^2 dg < \infty$.

It is easy to check that $\phi_f \in L^2(G_Q\backslash G_A,\psi)$ if $f \in S_k(N,\psi)$. Indeed Conditions 3.6(i)-(iii) and (v) imply that $|\phi_f(g)|^2$ is actually a function on $Z_A G_Q\backslash G_A/K_0^N K_\infty$. So since

(3.8) $$Z_A G_Q\backslash G_A/K_0^N K_\infty \cong \Gamma_0(N)\backslash SL(2,R)/SO(2)$$

(by (3.1) and (3.5)) we have

$$\int_{Z_A G_Q\backslash G_A} |\phi_f(g)|^2 dg = \int_{\Gamma_0(N)\backslash\{\mathrm{Im}(z)>0\}} |f(z)|^2 y^k \frac{dxdy}{y^2} < \infty,$$

which proves our claim.

DEFINITION 3.3. An *automorphic form on* GL(2) is any function $\phi$ on $G_A$ satisfying the following conditions:

(i) $\phi(\gamma g) = \phi(g)$ for all $\psi \in G_Q$;

(ii) for some grossencharacter $\psi$,

$$\phi(gz) = \phi(zg) = \psi(z)\phi(g) \text{ for all } z \in Z_A;$$

(3.9) (iii) $\phi$ is right $K = K_\infty \prod_{p<\infty} K_p$-finite;

(iv) as a function of $G_\infty$ alone, $\phi$ is smooth and $\mathfrak{z}$-finite (here $\mathfrak{z}$ denotes the center of the universal enveloping algebra of $G_\infty$); and

(v) $\phi$ is slowly increasing (in the sense of 3.6(vi)).

If, in addition to (i)-(v), the function $\phi$ satisfies the cuspidal condition 3.6(vii), namely

$$\int_{Q\backslash A} \phi\left(\begin{bmatrix} 1 & x \\ 0 & 1 \end{bmatrix} g\right) dx = 0 \text{ for a.e. g. ,}$$

then $\phi$ is a *cusp form on* GL(2) (or a $\psi$-cusp form, by 3.9(ii)). The space of all such cusp forms is denoted $A_0(\psi)$.

We have already remarked that the classical holomorphic cusp forms provide special examples of $\psi$-cusp forms in the above sense, and that these cusp forms are "square integrable on $Z_A G_Q \backslash G_A$." In general, one can show that *every $\psi$-cusp form belongs to* $L^2(G_Q \backslash G_A, \psi)$. (This statement, of course, is false for automorphic forms which are *not* cusp forms.)

REMARK 3.4. The subspace of $L^2(G_Q \backslash G_A, \psi)$ consisting of cuspidal functions shall be denoted $L_0^2(G_Q \backslash G_A, \psi)$ and referred to loosely as the space of $\psi$-cusp forms. This terminology is apt because $A_0(\psi)$ coincides with the dense subspace of $L_0^2(G_Q \backslash G_A, \psi)$ consisting of K-finite $\mathfrak{z}$-finite functions.

REMARK 3.5 (*Concerning the Fourier expansion of an arbitrary $\psi$-cusp form*). The Fourier expansion of a classical cusp form at a cusp s was introduced in Remark 1.3. The adelic formulation of this notion is quite simple and extends to arbitrary $\psi$-cusp forms $\phi$ (functions in $L_0^2(G_Q\backslash G_A,\psi)$) as follows.

For almost every g in $G_A$,

$$x \to \phi\left(\begin{bmatrix}1 & x\\ 0 & 1\end{bmatrix} g\right)$$

describes a square-integrable function of the compact group $Q\backslash A$. Recall that $A$ is a self-dual group of all of whose characters are of the form $\tau(bx)$, with b in $A$,

$$\tau(x) = \prod_{p\leq\infty} \tau_p(x_p)$$

$\tau_\infty(x_\infty) = e^{2\pi i x_\infty}$, and $\tau_p(x_p) = 1$ if and only if $x_p \in O_p$. The characters of $Q\backslash A$ are of the form $\tau_\xi(x) = \tau(\xi x)$, where $\xi$ is arbitrary in $Q$. Consequently we have the Fourier expansion

$$\phi\left(\begin{bmatrix}1 & x\\ 0 & 1\end{bmatrix} g\right) = \sum_{\xi\in Q} \phi_\xi(g)\tau(\xi x) \tag{3.10}$$

for each $\phi$. The $\xi$-th Fourier coefficient $\phi_\xi(g)$ is given by

$$a_\xi(g) = \phi_\xi(g) = \int_{Q\backslash A} \phi\left(\begin{bmatrix}1 & x\\ 0 & 1\end{bmatrix} g\right)\overline{\tau(\xi x)}\,dx \tag{3.11}$$

and clearly depends on g. *The cuspidal condition 3.6(vii), as expected, says that the zeroth Fourier coefficient* $a_0(g)$ *is zero for almost every* g. Also, for cusp forms of type $\phi_f$, $f \in S_k(N,\psi)$, this general Fourier expansion contains the expansions

$$f|[\sigma]_k(z) = \sum_{n=1}^{\infty} a_n e^{2\pi i n z/h} \tag{3.12}$$

introduced in Section 1. The verification of this fact is straightforward but amusing and therefore we include it below.

LEMMA 3.6. *Suppose* $f \in S_k(SL(2,\mathbf{Z}))$ *and* $\phi = \phi_f$. *Then for each* $y > 0$,

$$(3.13) \qquad \phi_\xi\left(\begin{bmatrix} y & 0 \\ 0 & 1 \end{bmatrix}\right) = \begin{cases} a_n e^{-2\pi n y} & \text{if } \xi = n \in \mathbf{Z} \\ 0 & \text{otherwise} . \end{cases}$$

*Consequently,*

$$\phi\left(\begin{bmatrix} 1 & x \\ 0 & 1 \end{bmatrix} g\right) = \sum_{\xi \in \mathbf{Q}} \phi_\xi(g)\tau(\xi x) = \sum_n a_n e^{2\pi i n z} = f(z)$$

*if* $g = \begin{bmatrix} y & 0 \\ 0 & 1 \end{bmatrix}$, $y > 0$, *and* $x \in \mathbf{R}$.

*Proof.* Suppose first that $\xi \notin \mathbf{Z}$. Then for some prime $p$, and some integer $m > 0$, $\xi = \alpha p^{-m}$, with $\alpha$ in $\mathbf{Q}$ relatively prime to $p$. But 3.9(iii) says that $\phi$ is right $K_0 = \prod_{p<\infty} GL(2,O_p)$ invariant. Therefore, since

$$\begin{bmatrix} 1 & p^{m-1} \\ 0 & 1 \end{bmatrix} = \Big(1_2, 1_2, \cdots, \underbrace{\begin{bmatrix} 1 & p^{m-1} \\ 0 & 1 \end{bmatrix}}_{\text{p-th place}}, 1_2, \cdots\Big)$$

belongs to $K_0$, it follows that

$$\phi_\xi\left(\begin{bmatrix} y & 0 \\ 0 & 1 \end{bmatrix}\right) = \int_{\mathbf{Q}\backslash\mathbf{A}} \phi\left(\begin{bmatrix} 1 & x \\ 0 & 1 \end{bmatrix}\begin{bmatrix} y & 0 \\ 0 & 1 \end{bmatrix}\begin{bmatrix} 1 & p^{m-1} \\ 0 & 1 \end{bmatrix}\right)\overline{\tau(\xi x)}\,dx$$

$$= \int_{\mathbf{Q}\backslash\mathbf{A}} \phi\left(\begin{bmatrix} 1 & x+p^{m-1} \\ 0 & 1 \end{bmatrix}\begin{bmatrix} y & 0 \\ 0 & 1 \end{bmatrix}\right)\overline{\tau(\xi x)}\,dx ,$$

where $x+p^{m-1}$ denotes the adele $(x,0,\cdots,0,p^{m-1},0,\cdots)$. So making the change of variable $x$ goes to $x-p^{m-1}$, we have

(3.14) $$\phi_\xi\left(\begin{bmatrix} y & 0 \\ 0 & 1 \end{bmatrix}\right) = \tau(\xi p^{m-1})\phi_\xi\left(\begin{bmatrix} y & 0 \\ 0 & 1 \end{bmatrix}\right).$$

But $\tau(\xi p^{m-1}) = \tau(a p^{-1}) = \tau(0,\cdots,0,ap^{-1},0,\cdots,0) \neq 1$ by our choice of $\tau$. Therefore (3.14) implies that $\phi_\xi\left(\begin{bmatrix} y & 0 \\ 0 & 1 \end{bmatrix}\right)$ must vanish identically.

On the other hand, suppose $\xi = n \in \mathbf{Z}$. Then

$$\begin{aligned}\phi_\xi\left(\begin{bmatrix} y & 0 \\ 0 & 1 \end{bmatrix}\right) &= \int_{Q\backslash A} \phi\left(\begin{bmatrix} y & x \\ 0 & 1 \end{bmatrix}\right)\overline{\tau(nx)}\,dx \\ &= \int_{Z\backslash R} \phi\left(\begin{bmatrix} y & x \\ 0 & 1 \end{bmatrix}\right)\overline{\tau(nx)}\,dx \\ &= \int_0^1 f(x+iy)\,e^{-2\pi inx}\,dx = a_n e^{-2\pi ny}\end{aligned}$$

and the proof of (3.13) is complete. The rest of the lemma follows from the fact that $\tau(x) = e^{2\pi inx}$ whenever $x$ is real. □

If $\Gamma = \Gamma_0(N)$ the relations between (3.10) and the classical Fourier expansions (1.6) are established similarly.

## B. Hecke Operators

We know from A that elements of $S_k(\Gamma)$ correspond to special functions in the Hilbert space $L^2(G_Q \backslash G_A/K_0^N)$. Our task now is to reinterpret the classical Hecke operators $T(p)$ in this adelic setting.

If $\Gamma = \Gamma_0(N)$, let $p$ denote any prime not dividing $N$. Then $K_p^N = K_p = GL(2, O_p)$. In $G_p$, consider the double coset

$$H_p = K_p\begin{bmatrix} p & 0 \\ 0 & 1 \end{bmatrix}K_p .$$

(This is the same thing as $K_p\begin{bmatrix} 1 & 0 \\ 0 & p \end{bmatrix}K_p$ since $w = \begin{bmatrix} 0 & 1 \\ -1 & 0 \end{bmatrix} \in K_p$ and $w\begin{bmatrix} p & 0 \\ 0 & 1 \end{bmatrix}w^{-1} = \begin{bmatrix} 1 & 0 \\ 0 & p \end{bmatrix}$.)

If $\phi$ is any function on $G_A$ right invariant by $K_p$, let $\tilde{T}(p)\phi$ denote the right convolution over $G_p$ of $\phi$ with the characteristic function of $H_p$. In particular, for each $\phi$ in $L^2(G_Q \backslash G_A / K_0^N)$,

$$\tilde{T}(p)\phi(g) = \int_{H_p} \phi(gh)\,dh \,. \tag{3.15}$$

Furthermore, since the elements of $G_\infty^+$ and $K_p$ commute (when viewed as adeles of GL(2) in the obvious way) it is easy to verify that each of the properties 3.6(i)-(vii) is preserved by $\tilde{T}(p)$ and, consequently, that the image of $S_k(\Gamma')$ in $L_0^2(G_Q \backslash G_A / K_0^N)$ is mapped into itself by $\tilde{T}(p)$.

LEMMA 3.7. *If* $\phi = \phi_f$, $f \in S_k(\Gamma)$, *then*

$$p^{\frac{k}{2}-1}\,\tilde{T}(p)\phi = \phi_{T(p)f} \,.$$

*Proof.* From the theory of "elementary divisors" in $G_p$, it follows that

$$H_p = \bigcup_{b=0}^{p-1} \begin{bmatrix} p & -b \\ 0 & 1 \end{bmatrix} K_p \cup \begin{bmatrix} 1 & 0 \\ 0 & p \end{bmatrix} K_p \,,$$

a disjoint union of right $K_p$-cosets. On the other hand, $\phi$ is right $K_p$-invariant. Consequently, it follows from (3.15) that

$$p^{\frac{k}{2}-1}\,\tilde{T}(p)\phi(g) = p^{\frac{k}{2}-1} \sum_{b=0}^{p-1} \phi\left(g\begin{bmatrix} p & -b \\ 0 & 1 \end{bmatrix}\right) + p^{\frac{k}{2}-1}\,\phi\left(g\begin{bmatrix} 1 & 0 \\ 0 & p \end{bmatrix}\right).$$

(The Haar measure on $G_p$ being normalized so that $K_p$ has mass 1.)

Now recall that $\phi(g) = \phi_f(g) = f(g_\infty(i))\,j(g_\infty, i)^{-k}$, if $g = \gamma g_\infty k_0$. Observe that

$$g_\infty k_0 \begin{bmatrix} p & -b \\ 0 & 1 \end{bmatrix} = \gamma' \begin{bmatrix} p^{-1} & bp^{-1} \\ 0 & 1 \end{bmatrix}_\infty g_\infty k_0' \,,$$

with

$$\gamma' = \left(\begin{bmatrix} p & -b \\ 0 & 1 \end{bmatrix}, \begin{bmatrix} p & -b \\ 0 & 1 \end{bmatrix}, \ldots, \begin{bmatrix} p & -b \\ 0 & 1 \end{bmatrix}, \ldots\right),$$

and

$$k_0' = k_0\left(1, \begin{bmatrix} p & -b \\ 0 & 1 \end{bmatrix}^{-1}, \begin{bmatrix} p & -b \\ 0 & 1 \end{bmatrix}^{-1}, \ldots, 1, \underset{\text{p-th place}}{\begin{bmatrix} p & -b \\ 0 & 1 \end{bmatrix}^{-1}}, \ldots\right).$$

Thus

$$\phi\left(g\begin{bmatrix} p & -b \\ 0 & 1 \end{bmatrix}\right) = f\left(\begin{bmatrix} p & -b \\ 0 & 1 \end{bmatrix}^{-1} z\right) p^{-k/2} = f\left(\frac{z+b}{p}\right) p^{-k/2},$$

similarly

$$\phi\left(g\begin{bmatrix} 1 & 0 \\ 0 & p \end{bmatrix}\right) = f\left(\begin{bmatrix} 1 & 0 \\ 0 & p \end{bmatrix}^{-1} z\right) = f(pz)p^{k/2},$$

and consequently

$$\begin{aligned} p^{k/2-1}\tilde{T}(p)\phi(g) &= p^{-1}\sum_{b=0}^{p-1} f\left(\frac{z+b}{p}\right) + p^{k-1}f(pz) \\ &= p^{k-1}\sum_{\substack{a>0 \\ ad=p}}\sum_{b=0}^{d-1} f\left(\frac{az+b}{d}\right)d^{-k} \\ &= \phi_{T(p)f}(g), \end{aligned}$$

as desired. □

REMARK 3.8 (*Why* GL(2) *and not* SL(2)?). There are numerous technical reasons why GL(2) (and not SL(2)) is taken as the starting point for the modern theory of automorphic forms. On the one hand the existence of a non-trivial center in GL(2) makes this group the natural habitat for classical cusp forms of weight $k$ and *non-trivial* conductor $\psi$. On the other there is the obvious fact that $\begin{bmatrix} p & 0 \\ 0 & 1 \end{bmatrix}$ does not belong to SL(2). This last fact gives rise to the following awkward situation.

Suppose we realize functions in $S_k(\Gamma)$ as special functions on $SL(2, A)$. Then for each prime $p$ it is natural to define the operator $\tilde{T}(p)\phi$ as convolution of $\phi$ with the characteristic function of

$$SL(2, O_p)\begin{bmatrix} p & 0 \\ 0 & p^{-1} \end{bmatrix} SL(2, O_p).$$

The problem is that *this* operator does *not* coincide with $T(p)$.

## C. Arbitrary Base Fields

Thus far we have regarded GL(2) as an algebraic group *defined over* $\mathbb{Q}$ since this base field clearly suffices for the classical theory as described in Section 1. However, other classical examples, such as Hilbert's modular forms, strongly suggest that we deal with automorphic forms on GL(2) defined over *arbitrary* global fields F.

So suppose first that F is an arbitrary algebraic number field, finite over $\mathbb{Q}$. Each place of F is denoted by v and the completion of F at v by $F_v$. As usual, if v is finite, $O_v$ denotes the ring integers of $F_v$, and $O_v^\times$ its group of units. The adeles of F are again denoted by A, but now we observe that

$$A^\times = F^\times(F_\infty^0)\left(\prod_{v \text{ finite}} O_v^\times\right), \tag{3.16}$$

*if and only if the class number of* F *is one.* (Here $F_\infty^0$ denotes the connected component of the archimedean component of F, the product of the completions $F_v$ as v ranges over the finite set of infinite places of F.)

Regarding GL(2) as an algebraic group *over* F we write $G_A$ for the corresponding adelized group $GL(2, A_F)$. Similarly $G_F = GL(2, F)$ and $G_v = GL(2, F_v)$. We then define a cusp form on GL(2) *over* F by suitably modifying the conditions of Definition 3.3. In particular, such a form satisfies the following conditions:

(i) $\phi(\gamma g) = \phi(g)$ for all $\gamma \in G_F$;

(ii) for some grossencharacter $\psi$ *of* F,

$$\phi(gz) = \psi(z)\phi(g) \quad \text{for all} \quad z \in Z_{A_F}; \tag{3.17}$$

(iii) $\phi(g)$ is right $K = K_\infty \prod_{v \text{ finite}} K_v$ finite if $K_v = GL(2, O_v)$;

(iv) viewed as a function of $G_\infty = \prod_{v \text{ infinite}} G_v$,

$\phi$ is smooth and $\mathfrak{z}_\infty$ finite ($\mathfrak{z}_\infty$ denotes the center of universal enveloping algebra of $G_\infty$);

(v) $\phi$ is slowly increasing; and

(vi) $\int_{F\backslash A} \phi\left(\begin{bmatrix} 1 & x \\ 0 & 1 \end{bmatrix} g\right) dx = 0$ for a.e.g.

As before, the space $A_0(\psi)$ of such forms is characterized as the dense subspace of $L_0^2(G_F \backslash G_A, \psi)$ consisting of K-finite $\mathfrak{z}_\infty$-finite functions.

REMARK 3.9. Suppose F is a function field in one variable over a finite field of constants. Then the notion of an automorphic cusp form on GL(2) over F is particularly simple since in this case there are no infinite places. In fact the space $A_0(\psi)$ is precisely the space of K-finite vectors in $L_0^2(G_F \backslash G_A, \psi)$ and conditions (iv) and (v) of (3.17) may be discarded. (Any K-finite cuspidal function on $G_F \backslash G_A$ *satisfying* 3.17(ii) is automatically *compactly supported* (modulo $Z_A$) if F is a function field.)

EXAMPLE 3.10 (Hilbert modular forms). Suppose $F = \mathbb{Q}(\sqrt{d})$ with $d > 0$. For simplicity suppose also that F has class number 1. Then it can be shown that

$$\text{(3.18)} \qquad GL(2, A_F) = G_F G_\infty^0 \prod_{v \text{ finite}} GL(2, O_v)$$

(using (3.16) and "strong approximation" for SL(2) over F). Therefore, since $G_\infty^0 = GL^+(2, \mathbb{R}) \times GL^+(2, \mathbb{R})$, and (essentially)

$$\text{(3.19)} \qquad (G_\infty^0 \Pi K_v) \cap G_F = SL(2, O_F) ,$$

where $O_F = \mathbb{Z}(\sqrt{d})$ denotes the ring of integers of F, we have

$$\text{(3.20)} \qquad Z_A G_F \backslash G_A / \Pi K_v \cong SL(2, O_F) \backslash SL(2, \mathbb{R}) \times SL(2, \mathbb{R}) .$$

Since $SL(2, \mathbb{R}) \times SL(2, \mathbb{R})$ acts naturally (and transitively) on $\{Im(z_1) > 0\} \times \{Im(z_2) > 0\}$, with isotropy group $SO(2) \times SO(2)$, we also have

$$Z_A G_F \backslash G_A / (\Pi K_v) K_\infty \cong SL(2, O_F) \backslash \{Im(z_1) > 0\} \times \{Im(z_2) > 0\} .$$

We recall that a Hilbert modular (cusp) form (of weight k) is a holomorphic function in $\{Im(z_1)>0\} \times \{Im(z_2)>0\}$ which is bounded and such that

$$f\left(\frac{az_1+b}{cz_1+d}, \frac{\bar{a}z_2+\bar{b}}{\bar{c}z_2+\bar{d}}\right) = (cz_1+d)^k (\bar{c}z_2+\bar{d})^k f(z_1, z_2)$$

whenever $\begin{bmatrix} a & b \\ c & d \end{bmatrix}$ belongs to $SL(2, O_F)$. Here $\bar{a}$ denotes the conjugate of a in $Q(\sqrt{d})$. Using (3.18) and (3.19) it is not difficult to show that any such $f(z_1, z_2)$ corresponds to a function $\phi(g)$ on $GL(2, A_F)$ satisfying the following conditions: $\phi(g) \in L_0^2(G_F \backslash G_A, 1)$, $\phi$ is right $\prod_{v \text{ finite}} K_v$-*invariant*, $\phi$ is an eigenfunction of $\mathfrak{z}_\infty$, and $\phi(gr(\theta_1) r(\theta_2)) = e^{-ik\theta_1} e^{-ik\theta_2} \phi(g)$ for all $(r(\phi_1), r(\phi_2)) \in K_\infty$. Thus $f(z_1, z_2)$ corresponds to a special automorphic cusp form on GL(2) over F.

## FURTHER NOTES AND REFERENCES

Suppose we are given an algebraic number field F and an algebraic group defined over F. If S is some finite set of places and $G_S$ is the S-component of $G_A$ then the problem of strong approximation asks whether $G_F G_S$ is dense in $G_A$. For $S = S_\infty$ (the set of infinite places of F) and $G = SL(2)$, the answer is yes. This means that for each $g_1$ in $SL(2, A(Q))$ we can find $\gamma_1$ in $SL(2, Q)$ such that the p-component of $\gamma_1^{-1} g_1$ belongs to $SL(2, O_p)$ for each p. On the other hand,

$$A_F^x / F^x F_\infty^0 \prod_{v \text{ finite}} O_v^x$$

is naturally isomorphic to the ideal class group of F for any number field F. Consequently, taking $F = Q$, (3.3) results. In particular, strong approximation for SL(2), together with the identity

$$g = \begin{pmatrix} r & 0 \\ 0 & 1 \end{pmatrix} g_1 \begin{pmatrix} u & 0 \\ 0 & 1 \end{pmatrix},$$

where g in $GL(2, A)$ is arbitrary, and $\det g = ru$, with $r \in Q^x$ and $u \in (Q_\infty)_+^x \prod_{p<\infty} O_p^x$, indeed implies (3.1).

The problem of strong approximation is discussed in [Kneser]. Some general features of automorphic forms on GL(2) over number fields and function fields are described in [Jacquet-Langlands], [Godement-Jacquet], and [Weil 2].

## §4. THE REPRESENTATIONS OF GL(2) OVER LOCAL AND GLOBAL FIELDS

In this section we shall describe some basic features of the theory of representations for GL(2) over local and global fields.

### A. The Archimedean Places

We deal first with $GL(2, \mathbf{R})$ since our primary interest is in automorphic forms defined *over* $\mathbf{Q}$. The theory for $GL(2, \mathbf{C})$ is simpler, requires little additional attention, and is included for the sake of completeness.

In Section 2 we have described the irreducible *unitary* representations of $SL(2, \mathbf{R})$. Our present point of departure is the well-known fact that any such representation $\pi$ is "admissible" in the following sense.

Consider the universal enveloping algebra of the complexification of the Lie algebra of $SL(2, \mathbf{R})$. If $\pi$ is realized on the Hilbert space H, and $\sigma$ is an irreducible representation of $K = SO(2)$, let $H(\sigma)$ denote the subspace of vectors v in H which transform according to $\sigma$. (Thus $\pi(K)v$ spans a finite-dimensional subspace of H equivalent to a finite number of copies of $\sigma$.) The algebraic direct sum

$$H^0 = \bigoplus_{\sigma} H(\sigma)$$

is dense in H and comprises the subspace of K-finite vectors. The crucial fact is that by differentiation $\pi$ induces a representation of the enveloping algebra on the space $H^0$. This representation is (algebraically) irreducible since the unitary representation $\pi$ is (topologically) irreducible. *It is "admissible" in the sense that its restriction to the Lie algebra of* K *decomposes into finite-dimensional representations with finite multiplicities.* (Equivalently, the dimension of $H(\sigma)$ is finite for each $\sigma$.)

Following Harish-Chandra and Jacquet-Langlands we henceforth focus our attention not on irreducible *unitary* representations of $G = GL(2, \mathbf{R})$ but rather on irreducible *admissible* representations of an appropriate group algebra of G.

DEFINITION 4.1. Let $\mathfrak{G}$ denote the universal enveloping algebra of the complexification of the Lie algebra of G. Let $\varepsilon_-$ denote the Dirac measure at $\begin{bmatrix} -1 & 0 \\ 0 & 1 \end{bmatrix}$. Then the *Hecke group algebra of* G is by definition the direct sum

$$\mathcal{H}(G) = \mathfrak{G} \oplus (\varepsilon_-) * \mathfrak{G} .$$

Observe that $\mathcal{H}(G)$ is an algebra (under convolution product) of distributions supported in the subgroup $\left\{\begin{bmatrix} \pm 1 & 0 \\ 0 & 1 \end{bmatrix}\right\}$ of G. It is the simplest substitute for the enveloping algebra of G which takes into account the fact that G has *two* connected components.

DEFINITION 4.2. Suppose $\pi$ is a representation of the algebra $\mathcal{H}(G)$ on a complex vector space V. Then $\pi$ is said to be *admissible* if its restriction to the Lie algebra of $K = 0(2, \mathbf{R})$ decomposes into finite-dimensional representations with finite multiplicities.

REMARK 4.3. Almost all the irreducible representations of $K = 0(2, \mathbf{R})$ are two dimensional. Suppose $\sigma$ is any such representation of (the Lie algebra of) K and $V(\sigma)$ denotes the subspace of vectors in V which transform according to some multiple of $\sigma$. Then admissibility of $(\pi, V)$ amounts to the assertion that

$$V = \bigoplus_{\sigma} V(\sigma) \qquad (4.1)$$

(algebraic sum)

with each $V(\sigma)$ *finite dimensional.*

Our task now is to describe all the *irreducible admissible* representations up to equivalence. This task is simplified by the fact that all such representations are known to be subquotients of the following basic representations of $\mathcal{H}(G)$, the so-called *principal* series *representations.*

Let $\mu_1, \mu_2$ denote any quasi (i.e. not necessarily unitary) characters of $\mathbb{R}^\times$. Let $\mathcal{H}(\mu_1, \mu_2)$ denote the vector space of functions $\phi(g)$ on G which are right K-finite (cf. the definition in Section 2A) and such that

(4.2) $$\phi\left(\begin{bmatrix} t_1 & * \\ 0 & t_2 \end{bmatrix} g\right) = \mu_1(t_1)\mu_2(t_2)\left|\frac{t_1}{t_2}\right|^{1/2} \phi(g)$$

for all $t_1, t_2 \in \mathbb{R}^\times$. According to the Iwasawa decomposition,

(4.3) $$G = NA\, SO(2)$$

(where $A = \left\{\begin{bmatrix} t_1 & 0 \\ 0 & t_2 \end{bmatrix}\right\}$). Therefore, each $\phi$ in $\mathcal{H}(\mu_1, \mu_2)$ is completely determined by its restriction to $SO(2,\mathbb{R})$ (and in particular is infinitely differentiable since this restriction must be a trigonometric polynomial).

If X belongs to the Lie algebra we define (as usual)

(4.4) $$\phi * X(g) = \frac{d}{dt}\phi(g \exp(-tX))\big|_{t=0}$$

and let $\check{X}$ denote $-X$. By $\rho(\mu_1, \mu_2)$ we denote the representation of $\mathcal{H}(G)$ on $\mathcal{H}(\mu_1, \mu_2)$ determined by

(4.5) $$\rho(\mu_1, \mu_2)(X)\phi = \phi * \check{X}\ .$$

This representation is the lift to the enveloping algebra of the representation of G induced from the character

$$\begin{bmatrix} t_1 & * \\ 0 & t_2 \end{bmatrix} \to \mu_1(t)\mu_2(t)$$

of B. It is admissible precisely because each $\phi$ in $\mathcal{H}(\mu_1, \mu_2)$ restricts to a trigonometric polynomial on $SO(2,\mathbb{R})$. What is remarkable is that *every irreducible admissible representation of* $\mathcal{H}(G)$ *is a subquotient of some such* $\rho(\mu_1, \mu_2)$. Therefore the classification of irreducible admissible representations is reduced to the study of how (and when) these $\rho(\mu_1, \mu_2)$ decompose.

To analyze the reducibility of $\rho(\mu_1, \mu_2)$ one can compute the action of certain Lie algebra elements on convenient basis elements of $\mathcal{H}(\mu_1, \mu_2)$. To this end, write $\mu_i(t) = |t|^{s_i}[\operatorname{sgn}(t)]^{m_i}$, where $m_i = 0$ or $1$, and $s_i \in \mathbf{C}$. Then

$$\mu(t) = \mu_1\mu_2^{-1}(t) = |t|^s[\operatorname{sgn}(t)]^m \tag{4.6}$$

with $s = s_1 - s_2$ and $m = m_1 - m_2$. For each $n \equiv m \pmod 2$ the function

$$\phi_n\left(\begin{bmatrix} t_1 & * \\ 0 & t_2 \end{bmatrix}\begin{bmatrix} \cos\theta & -\sin\theta \\ \sin\theta & \cos\theta \end{bmatrix}\right) = \mu_1(t_1)\mu_2(t_2)\left|\frac{t_1}{t_2}\right|^{1/2} e^{-in\theta} \tag{4.7}$$

belongs to $\mathcal{B}(\mu_1, \mu_2)$ and the set $\{\phi_n\}$ is obviously a basis for $\mathcal{B}(\mu_1, \mu_2)$. The relevant elements of $\mathfrak{G}$ are

$$U = \begin{bmatrix} 0 & 1 \\ -1 & 0 \end{bmatrix} \qquad J = \begin{bmatrix} 1 & 0 \\ 0 & 1 \end{bmatrix}$$

$$V_+ = \begin{bmatrix} 1 & i \\ i & -1 \end{bmatrix} \qquad V_- = \begin{bmatrix} 1 & -i \\ -i & -1 \end{bmatrix} \qquad Z = \begin{bmatrix} 1 & 0 \\ 0 & -1 \end{bmatrix}$$

$$X_+ = \begin{bmatrix} 0 & 1 \\ 0 & 0 \end{bmatrix} \qquad X_- = \begin{bmatrix} 0 & 0 \\ 1 & 0 \end{bmatrix}$$

as well as the Casimir operator

$$D = X_+X_- + X_-X_+ + \frac{Z^2}{2}$$

which belongs to (the center of) $\mathfrak{G}$. Writing $\rho$ for $\rho(\mu_1, \mu_2)$, we have

(i) $\rho(U)\phi_n = in\phi_n$

(ii) $\rho(V_+)\phi_n = (s+1+n)\phi_{n+2}$

(iii) $\rho(V_-)\phi_n = (s+1-n)\phi_{n-2}$

(iv) $\rho(\varepsilon_-)\phi_n = (-1)^{m_1}\phi_{-n}$

(v) $\rho(D)\phi_n = \dfrac{s^2-1}{2}\phi_n$ and

(vi) $\rho(J)\phi_n = (s_1+s_2)\phi_n$ .

From this it follows that

$$\rho(V_+)^p\phi_n = (s+1+n)\cdots(s+2p-1+n)\phi_{n+2p}$$

and

$$\rho(V_-)^p\phi_n = (s+1-n)\cdots(s+2p-1-n)\phi_{n-2p} \,.$$

Now the elements $U, Z, V_+$ and $V_-$ form a $\mathbb{C}$-basis for the Lie algebra of $G$ and the eigenvectors of $\rho(U)$ span $\mathcal{H}(\mu_1, \mu_2)$. Consequently the above computations show that $\rho(\mu_1, \mu_2)$ is reducible if and only if $s-m$ is an odd integer, i.e. if and only if $\mu(t) = t^p \operatorname{sgn}(t)$ for some integer $p$. More precisely, we have:

THEOREM 4.4.

(a) *If* $\mu_1\mu_2^{-1}$ *is not of the form* $t \to t^p \operatorname{sgn}(t)$, *with* $p$ *a non-zero integer, then* $\mathcal{H}(\mu_1, \mu_2)$ *is irreducible under the action of* $\mathcal{H}(G)$;

(b) *If* $\mu_1\mu_2^{-1}(t) = t^p \operatorname{sgn}(t)$, *with* $p > 0$, *then* $\mathcal{H}(\mu_1, \mu_2)$ *contains exactly one invariant subspace, namely,*

$$\mathcal{H}^s(\mu_1, \mu_2) = \{\ldots, \phi_{-p-3}, \phi_{-p-1}, \phi_{p+1}, \phi_{p+3}, \ldots\}$$

*and the quotient* $\mathcal{H}^f(\mu_1, \mu_2) = \mathcal{H}/\mathcal{H}^s(\mu_1, \mu_2)$ *is finite dimensional;*

(c) *If* $\mu(t) = t^p \operatorname{sgn}(t)$ *with* $p < 0$ *then the only invariant subspace of* $\mathcal{H}(\mu_1, \mu_2)$ *is*

$$\mathcal{H}^f(\mu_1, \mu_2) = \{\phi_{p+1}, \phi_{p+3}, \ldots, \phi_{-p-3}, \phi_{-p-1}\} \,.$$

The representation $\rho(\mu_1, \mu_2)$ will be denoted by $\pi(\mu_1, \mu_2)$ if it is irreducible; in case it is not, the obvious representation on the finite dimensional space $\mathcal{H}^f(\mu_1, \mu_2)$ will still be denoted by $\pi(\mu_1, \mu_2)$. The representation on the *infinite* dimensional subspace (or quotient) $\mathcal{H}^s(\mu_1, \mu_2)$ will be denoted by $\sigma(\mu_1, \mu_2)$ and viewed as a member of the *discrete series* for $G$. It is defined only when $\mu(t) = t^p \operatorname{sgn}(t)$ for some non-zero integer $p$. The representations $\pi(\mu_1, \mu_2)$ exhaust the so-called *principal series* for $G$.

THEOREM 4.5. *Every irreducible admissible representation of* $\mathcal{H}(G)$ *is either a* $\pi(\mu_1, \mu_2)$ *or a* $\sigma(\mu_1, \mu_2)$; *the only equivalences between these representations are the following*:

$$\pi(\mu_1, \mu_2) \approx \pi(\mu_2, \mu_1)$$

*and*

$$\sigma(\mu_1, \mu_2) \approx \sigma(\mu_2, \mu_1) \approx \sigma(\mu_1\eta, \mu_2\eta) \approx \sigma(\mu_2\eta, \mu_1\eta) ,$$

*where* $\eta(t) = \mathrm{sgn}(t)$.

REMARK 4.6. The irreducible admissible representations of $\mathcal{H}(G)$ are often indexed by two particularly convenient parameters. For the (infinite dimensional) principal series representation $\pi(\mu_1, \mu_2)$ these parameters are $s = s_1 - s_2$ and $t = s_1 + s_2$ (cf. 4.5). Note then that two principal series representations $\pi$ and $\pi'$ will share the same eigenvalue for the Casimir operator if and only if $|s| = |s'|$. (They will share the same central character if and only if $t = t'$). If $\pi$ is equivalent to the discrete series representation $\sigma(\mu_1, \mu_2)$ then the convenient parameters for $\pi$ are (again) $p = s_1 - s_2$ and $t = s_1 + s_2$ (if $\mu_i(t) = |t|^{s_i}[\mathrm{sgn}(t)]^{m_i}$). The significance of $p$ is that the space of $\pi(p, t)$ will contain the functions $\phi_{p+1}, \phi_{p+3}, \cdots$, but not the function $\phi_{p-1}$. Consequently $\phi_{p+1}$ may be viewed as *the lowest weight vector for* $\pi(p, t)$.

REMARK 4.7. It can be shown that a given principal series representation $\pi(\mu_1, \mu_2)$ of $\mathcal{H}(G)$ corresponds to a *unitary* representation of $G$ if and only if either both $\mu_1$ and $\mu_2$ are unitary, i.e. the parameters $s$ and $t$ just introduced are both pure imaginary, or just $t$ is imaginary and $s = s_1 - s_2$ is real, non-zero, and between $-1$ and $1$. In accordance with the terminology already introduced for $SL(2, \mathbf{R})$, $\pi(\mu_1, \mu_2)$ is called a *continuous series* representation when this first possibility obtains, and a *complementary series* representation when the second does.

The discrete series representations $\sigma(\mu_1, \mu_2)$ are similarly "unitarizable" if and only if their central character is unitary, i.e. if and only if $t = s_1 + s_2$ is pure imaginary (cf. 4.7(vi)).

REMARK 4.8. (GL(2, C)).

Since GL(2, C) is connected it is reasonable to expect that in this case the correspondence between "nice" representations of G and "nice" representations of the Lie algebra of G will be one-to-one. Therefore for GL(2, C) the group algebra $\mathcal{H}(G)$ is defined to be precisely the enveloping algebra of the (complexification of the) Lie algebra of G. The "principal series" $\rho(\mu_1, \mu_2)$ are introduced in the obvious way for each pair of quasi-characters $\mu_1, \mu_2$ of $C^x$ and again it is the case that every irreducible *admissible* representation of $\mathcal{H}(G)$ is a subquotient of some $\rho(\mu_1, \mu_2)$. Furthermore, $\rho(\mu_1, \mu_2)$ is irreducible *except* when $\mu_1\mu_2^{-1}(z) = z^p\bar{z}^q$, with $p, q, \in Z$ and $pq > 0$, in which case $\mathcal{H}(\mu_1, \mu_2)$ "contains" exactly one infinite dimensional irreducible subquotient. *However, in contrast to the real case, this irreducible subquotient is actually equivalent to some irreducible* $\rho(\mu'_1, \mu'_2)$ *for appropriate* $\mu'_1, \mu'_2$! In particular, there are no discrete series for GL(2, C).

## B. The p-adic Theory

Our purpose is to describe the classification of all irreducible *admissible* representations of

$$G = GL(2, F)$$

when F is a non-archimedean locally compact field. We also shall collect some basic facts concerning these representations. These facts and others play a crucial role in the theory of automorphic forms.

### 1. *Admissibility*

We start with some notation. Throughout this section, F will be a finite extension of the p-adic number field *or* a field of formal power series in one variable over a finite field. The symbol $O_F$ will denote the ring of integers of F. The absolute value on F is defined by the relation $d(ax) = |a|\,dx$ where dx is an invariant measure on the additive group of F. The prime ideal of $O_F$, defined by $|a| < 1$, will be denoted by $P_F$. It is generated (say) by $\varpi$, and $O_F/P_F$ is a finite field with q elements,

$q$ some power of $p$. The standard maximal compact (open) subgroup of $G$ is

$$K = GL(2, O_F) .$$

Now $G$ has no enveloping algebra of differential operators but there is a natural group algebra for $G$ which is fundamental for its representation theory, namely the *(Hecke) group algebra* $\mathcal{H}(G)$ consisting of all locally constant compactly supported functions on $G$. (This is an algebra for the group convolution product

$$f * g(x) = \int_G f(xy^{-1})g(y)d^*y ,$$

where $d^*y$ denotes the Haar measure for $G$ which assigns the measure 1 to $K$.)

Following Jacquet and Langlands, we say that a representation $\pi$ of $\mathcal{H}(G)$ on a complex vector space $V$ is *admissible* if for every $v$ in $V$ there is an $f$ in $\mathcal{H}(G)$ such that $\pi(f)v = v$, and if every $\pi(f)$ maps $V$ onto a finite dimensional space. This definition of admissibility is motivated by the following considerations.

Suppose $\pi$ is an irreducible *unitary* representation *of* $G$ on some Hilbert space $H$. Then $\pi$ induces a representation of $\mathcal{H}(G)$ through the formula

$$\pi(f) = \int_G f(g)\pi(g)d^*g ,$$

and, as in the real case, $\pi(f)$ defines a representation of $\mathcal{H}(G)$ *in the subspace of* K-*finite functions of* $H$ which is admissible in the above sense. However, *in contrast to the real case,* $\pi(g)$ *itself acts in the space of* K-*finite vectors.* Therefore it should not seem surprising that *for p-adic groups (and admissible representations) the correspondence between representations of* $G$ *and representations of the group algebra of* $G$ *is completely transparent.* To see this, we make the following definition.

DEFINITION 4.9. Suppose $\pi$ is a representation of $G$ on a complex vector space $V$ (the space of K-finite vectors of some unitary G-space, for example). Then $\pi$ is said to be *admissible* if (i) the stabilizer in G of each v in V is an open subgroup of K, and (ii) the subspace of V fixed by any open subgroup of K is finite-dimensional.

REMARK 4.10. In Definition 4.9 *no* topology is imposed on the space V and so Condition (i) is meant as a substitute for the usual continuity condition for $\pi$. Condition (ii) is really the crucial assumption of admissibility. It implies that each operator

$$\pi(f) = \int_G f(g)\pi(g)d^*g$$

is of finite rank and therefore that *the character of* $\pi$, defined by

$$\theta_\pi(f) = \text{Trace } \pi(f) ,$$

always exists (at least as a "distribution" on G).

LEMMA 4.11. *If* $\pi(g)$ *is an admissible representation of* G, *then* $\pi(f)$ *defines an admissible representation of* $\mathcal{H}(G)$. *More significantly, if* $\pi(f)$ *is an admissible representation of* $\mathcal{H}(G)$, *then there exists an admissible representation* $\pi$ *of* G *such that*

$$\pi(f) = \int_G f(g)\pi(g)d^*g .$$

The proof of this lemma is tedious but straightforward. Since it is to be found in Section 2 of [Jacquet-Langlands] we simply content ourselves with the observation that for each v in V, the integral

$$\pi(f)v = \int_G f(g)\pi(g)vd^*g$$

is actually a finite sum. Indeed, since f belongs to $\mathcal{H}(G)$, Condition (i) of Definition 4.9 implies that $f(g)\pi(g)v$ takes on only finitely many values.

REMARK 4.12. The correspondence (established in Lemma 4.11) between admissible representations of G and $\mathcal{H}(G)$ is more than simply one-to-one. For if $V_1$ is a subspace of V invariant for G it is also invariant for $\mathcal{H}(G)$, and vice versa. Similarly, an operator in V commutes with the action of G if and only if it commutes with the action of $\mathcal{H}(G)$.

Henceforth we shall deal exclusively with irreducible admissible representations of $\mathcal{H}(G)$. Before describing the classification of such representations we offer yet another useful definition of admissibility.

DEFINITION 4.13. Suppose $\pi$ is a representation of G on a complex vector space V. Then the restriction of $\pi$ to K contains any given irreducible representation of K at most finitely many times. Equivalently, if $V(\sigma)$ denotes the subspace of v in V which transform according to the continuous irreducible representation $\sigma$ of K, then

(i) $V = \bigoplus_{\sigma} V(\sigma)$
(algebraic)

and

(ii) the dimension of each $V(\sigma)$ is finite.

This definition is entirely analogous to the usual definition for Lie groups. Its equivalence with Definition 4.9 seems never to have been verified in print and therefore we write out the details of this argument below:

LEMMA 4.14. *Definitions 4.9 and 4.13 are equivalent.*

*Proof.* Suppose first that $\pi$ is admissible in the sense of Definition 4.9. To prove that

$$V = \bigoplus_{\sigma} V(\sigma) \tag{4.8}$$

we consider the stabilizer U in K of any vector v in V. By Condition (i) of Definition 4.9 this U is open, and therefore we may find

$k_1, \cdots, k_N$ such that $K = \bigcup_{i=1}^{N} k_i U$. Since this implies that the space of translates of v under K is finite-dimensional, (4.8) follows by the arbitrariness of v. To prove that each $V(\sigma)$ is finite-dimensional we recall that each continuous irreducible representation of K is finite-dimensional. Consequently, there is an open subgroup $K^*$ of K such that $\sigma(k)$ is the identity operator for each $k \in K^*$. (Indeed, if $\{v_1, \cdots, v_d\}$ is any basis for the space of $\sigma$ in V, $K^*$ may be taken to be the intersection of the stability groups of these $v_i$.) But this implies that $V(\sigma)$ is contained in the space of fixed vectors for $K^*$. Therefore $V(\sigma)$ is finite-dimensional by Condition (ii) of Definition 4.9.

Now suppose that $\pi$ is admissible in the sense of Definition 4.13 and fix any v in V. By Condition (i) and (ii) of Definition 4.13 we may assume (without any loss of generality) that v belongs to an irreducible subspace of some finite-dimensional space $V(\sigma)$. But then the stability subgroup of v contains some open subgroup of K by the continuity of $\sigma$. Therefore it remains only to verify that Condition (ii) of Definition 4.9 holds.

To this end let U denote an arbitrary open subgroup of K and $V_U$ the corresponding subspace of V fixed by U. Without loss of generality we may assume that U is normal in K. To prove that $V_U$ is finite-dimensional it will suffice to prove that $V_U$ is contained in the direct sum of a *finite* number of spaces $V(\sigma_i)$ (since each of these spaces is finite-dimensional by Condition (ii) of Definition 4.13). But now we claim that only *finitely* many representations $\sigma$ contain the identity representation upon restriction to U and that the direct sum of the spaces in V fixed by *these* representations contains $V_U$. Indeed if $\sigma_i$ is such a representation, and w is any U-invariant vector in the space of $\sigma_i$, then $\pi(K)w$ is also U-invariant (since $\pi(u)\pi(k)w = \pi(k)w$ if and only if $\pi(k^{-1})\pi(u)\pi(k)w = w$, an obvious identity by the normality of U in K). Therefore there is a one-to-one correspondence between such $\sigma$ and equivalence class of irreducible representations of the *finite* group K/U (on $V(\sigma) \cap V_U$). From this our claim (and hence the lemma) follows. □

REMARK 4.15 (*Schur's Lemma*).

Schur's Lemma holds for all irreducible admissible representations, unitary or not. Indeed, any linear operator A in V which commutes with $\pi$ also operates in every *finite* dimensional subspace $\pi(f)(V)$ of V. Therefore A must have at least one eigenvector $v_0$. Since $\pi(g)v_0$ is again an eigenvector for A (for any $g \in G$) it follows from the irreducibility of $\pi$ that A is a scalar.

One consequence of Schur's Lemma is that any irreducible admissible representation of G defines a character of the center Z of G (the so-called *central character*). Another is that *every finite-dimensional irreducible admissible representation of* GL(2) (*over a* p-*adic field*) *is one dimensional* and of the form $\chi(\det g)$ for some quasi-character $\chi$ of $F^x$. Indeed, if the space of $\pi$ is finite-dimensional, its kernel is an *open* (hence non-trivial) normal subgroup of G. Now this kernel certainly contains

$$\begin{bmatrix} 1 & x \\ 0 & 1 \end{bmatrix} = \begin{bmatrix} a^{-1} & 0 \\ 0 & 1 \end{bmatrix} \begin{bmatrix} 1 & ax \\ 0 & 1 \end{bmatrix} \begin{bmatrix} a & 0 \\ 0 & 1 \end{bmatrix}$$

and $\begin{bmatrix} 1 & 0 \\ x & 1 \end{bmatrix}$ since $|ax|$ can be made arbitrarily small for a sufficiently small. Therefore the kernel of $\pi$ contains the group generated by the elements $\begin{bmatrix} 1 & x \\ 0 & 1 \end{bmatrix}$ and $\begin{bmatrix} 1 & 0 \\ x & 1 \end{bmatrix}$, namely SL(2, F)! So since $G = \left\{\begin{bmatrix} \det g & 0 \\ 0 & 1 \end{bmatrix}\right\} SL(2,F)$ our claim follows from Schur's Lemma.

2. *Classification of Admissible Representations*

The main classification result is that every irreducible admissible representation $\pi$ of G is a subrepresentation of some principal series representation *unless* $\pi$ is a so-called supercuspidal representation characterized by the fact that its matrix coefficients are compactly supported modulo the center. This is to be contrasted with the real case where no such supercuspidal representations exist.

To describe the principal series, let $\mu_1, \mu_2$ be any two quasi-characters of $F^x$ and let $\mathcal{H}(\mu_1, \mu_2)$ denote the space of all locally constant functions $\phi$ on G such that

(4.9)
$$\phi\left(\begin{bmatrix} t_1 & * \\ 0 & t_2 \end{bmatrix} g\right) = \mu_1(t_1)\mu_2(t_2)|t_1/t_2|^{1/2}\phi(g)$$

for all $t_1, t_2 \in F^{\times}$. The group G acts on $\mathcal{H}(\mu_1, \mu_2)$ through right translations and the resulting representation of G is called a *principal series representation* (at least when it is irreducible) and denoted by $\rho(\mu_1, \mu_2)$. Each such representation is clearly admissible. Indeed the Iwasawa decomposition

(4.10)
$$G = BK$$

implies that functions in $\mathcal{H}(\mu_1, \mu_2)$ are completely determined by their restriction to K. In fact $\mathcal{H}(\mu_1, \mu_2)$ is naturally isomorphic to the space of locally constant functions on K such that

$$\phi\left(\begin{bmatrix} u_1 & x \\ 0 & u_2 \end{bmatrix} k\right) = \mu_1(u_1)\mu_2(u_2)\phi(k)$$

for all $u_1, u_2 \in O_F^{\times}$ and $x \in O_F$. Therefore, if U is any open subgroup of K, the restriction of a function invariant under U is a function on the *finite* set K/U. Thus the space of all such functions is finite-dimensional (cf. Definition 4.9).

It is important to point out that $\rho(\mu_1, \mu_2)$ is nothing more or less than the representation of G induced from the one-dimensional representation

$$\begin{bmatrix} t_1 & x \\ 0 & t_2 \end{bmatrix} \to \mu_1(t_1)\mu_2(t_2)$$

of $B = \left\{\begin{bmatrix} t_1 & x \\ 0 & t_2 \end{bmatrix} : t_i \in F^{\times}, x \in F\right\}$.

DEFINITION 4.16. Suppose $\pi$ is any admissible representation of G on V. Let $\tilde{V}$ denote the subspace of those $v^*$ in $V^*$ (the algebraic dual of V) which are invariant under some open subgroup of G. Then the *contragredient of* $\pi$, denoted by $\tilde{\pi}(g)$, is the restriction of ${}^t\pi(g^{-1})$ to $\tilde{V}$.

Because

$$\tilde{V} = \bigoplus_{\sigma} V(\sigma)^*$$

it is clear that $\tilde{\pi}$ is admissible as soon as $\pi$ is. Furthermore, if $V_1$ is an invariant subspace of V for $\pi$, then its "orthogonal complement"

$$(V_1)^{\perp} = \{v^* \in V : v^*(v) = 0 \text{ for all } v \in V_1\}$$

is invariant for $\check{\pi}$ and $(V_1^{\perp})^{\perp} = V_1$. Therefore there is a one-to-one correspondence between invariant subspaces of V and $\check{V}$. Note that $(\tilde{\pi}(g)\tilde{v}, \pi(g)v) = (\tilde{v}, v)$ if $(\cdot,\cdot)$ denotes the duality between V and $\tilde{V}$. Note also that $(\tilde{\pi})^{\sim} = \pi$. This notion of contragredience is useful throughout the theory of automorphic forms and group representations and in particular helps resolve the question of reducibility for $\rho(\mu_1, \mu_2)$.

LEMMA 4.17. *The contragredient of* $\rho(\mu_1, \mu_2)$ *is* $\rho(\mu_1^{-1}, \mu_2^{-1})$.

This lemma results from a straightforward application of the definition of induced representation and the fact that the (one-dimensional) representations inducing $\rho(\mu_1, \mu_2)$ and $\rho(\mu_1^{-1}, \mu_2^{-1})$ are obviously contragredient to one another.

THEOREM 4.18. *The representation* $\rho(\mu_1, \mu_2)$ *is irreducible except when* $\mu(x) = \mu_1\mu_2^{-1}(x) = |x|$ *or* $|x|^{-1}$. *If* $\mu(x) = |x|^{-1}$, *then* $\rho(\mu_1, \mu_2)$ *contains a one-dimensional invariant subspace and the representation induced on the resulting factor space is irreducible. If* $\mu(x) = |x|$, *then* $\rho(\mu_1, \mu_2)$ *contains an irreducibly invariant subspace of codimension one. (The irreducible representations* $\rho(\mu_1, \mu_2)$ *are denoted by* $\pi(\mu_1, \mu_2)$ *and called principal series representations; if* $\mu_1\mu_2^{-1}(x) = |x|$ *or* $|x|^{-1}$ *the resulting irreducible subquotients of* $\rho(\mu_1, \mu_2)$ *are again denoted by* $\pi(\mu_1, \mu_2)$ *but are called special representations.)*

*Proof* (Sketch). Roughly speaking, the idea is to analyze explicitly the "Fourier transform" description of a certain "non-compact realization" of the induced representation $\rho(\mu_1, \mu_2)$.

More precisely, recall the *Bruhat decomposition* for G. This says that

$$G = B \cup B\begin{bmatrix} 0 & -1 \\ 1 & 0 \end{bmatrix}N \,. \tag{4.11}$$

(In particular, the "big cell" $B\begin{bmatrix} 0 & -1 \\ 1 & 0 \end{bmatrix}N$ is everywhere dense in G.) Now let $w = \begin{bmatrix} 0 & 1 \\ -1 & 0 \end{bmatrix}$. Then (4.11) together with the transforming property (4.9) implies that each $\phi \in \mathcal{H}(\mu_1, \mu_2)$ is completely determined by its restriction to $w^{-1}N$. Therefore $\mathcal{H}(\mu_1, \mu_2)$ is naturally isomorphic to the space $F(\mu_1, \mu_2)$ consisting of locally constant functions

$$\emptyset(x) = \phi\left(w^{-1}\begin{bmatrix} 1 & x \\ 0 & 1 \end{bmatrix}\right)$$

on F with the property that $\mu(x)|x|\,\emptyset(x)$ is constant for $|x|$ large.

Now for each $\emptyset \in F(\mu_1, \mu_2)$ consider the "twisted" Fourier transform

$$\hat{\emptyset}(x) = \mu_2(x)|x|^{1/2}\int_F \emptyset(y)\overline{\tau(xy)}\,dy \,. \tag{4.12}$$

Here $\tau(x)$ denotes any fixed non-trivial character of the additive group of F. Some p-adic analysis shows that the integral in (4.12) converges in some appropriate sense and that the resulting map

$$\phi \to \hat{\emptyset}_\phi \tag{4.13}$$

is injective *unless* $\mu(x) = |x|^{-1}$ (in which case its kernel is the one-dimensional subspace of $\mathcal{H}(\mu_1, \mu_2)$ spanned by $\phi_0(g) = \mu_1(\det g)|\det g|^{1/2}$). On the other hand, the image space of the map (4.13), call it $K(\mu_1, \mu_2)$, consists of locally constant functions on $F^x$ which vanish when $|x|$ is large, and a trivial formal computation shows that

$$\hat{\emptyset}_{\phi_*}(x) = \tau(bx)\hat{\emptyset}_\phi(ax) \tag{4.14}$$

if

$$\phi_*(g) = \left(\rho(\mu_1,\mu_2)\left(\begin{bmatrix} a & b \\ 0 & 1\end{bmatrix}\right)\phi\right)(g) .$$

That is, on $K(\mu_1,\mu_2)$ the restriction of $\rho(\mu_1,\mu_2)$ to the subgroup $\left\{\begin{bmatrix} a & b \\ 0 & 1\end{bmatrix}\right\}$ has a particularly simple description.

Suppose, finally, that $\mathcal{H}(\mu_1,\mu_2)$ contains some non-trivial invariant subspace V. The corresponding invariant subspace of $K(\mu_1,\mu_2)$ will then contain some non-zero function $\hat{\emptyset}(x)$, hence $\hat{\emptyset}(x)-\tau(bx)\hat{\emptyset}(x)$ for all $b \in F$ (by (4.14)), hence some non-zero function in $S(F^*)$, the *Schwartz-Bruhat space* of locally constant compactly supported functions on $F^{\times}$. (For each b in F the function $1-\tau(bx)$ vanishes in some neighborhood of the origin.) But now $S(F^{\times})$ can easily be shown to be irreducible under the action of the operators $\hat{\emptyset}(x) \to \tau(bx)\hat{\emptyset}(ax)$. Therefore any non-trivial invariant subspace of $K(\mu_1,\mu_2)$ contains the function $\hat{\emptyset}(x) - \tau(bx)\hat{\emptyset}(x)$ *for every* $\hat{\emptyset}$ *in* $K(\mu_1,\mu_2)$, or equivalently, the invariant subspace V of $\mathcal{H}(\mu_1,\mu_2)$ contains $\phi-\rho(\mu_1,\mu_2)\left(\begin{bmatrix} 1 & b \\ 0 & 1\end{bmatrix}\right)\phi$ *for all* $\phi$ *in* $\mathcal{H}(\mu_1,\mu_2)$ (assuming $\mu(x) \neq |x|^{-1}$). So by Lemma 4.17 the subspace $V^{\perp}$ orthogonal to V must consist of functions $\psi(g)$ in $\mathcal{H}(\mu_1^{-1},\mu_2^{-1})$ which are invariant under all $\rho\left(\begin{bmatrix} 1 & b \\ 0 & 1\end{bmatrix}\right)$, $b \in F$. (Indeed $\psi \in V^{\perp}$ implies $0 = \langle\phi,\psi\rangle - \langle\rho(\mu_1,\mu_2)\left(\begin{bmatrix} 1 & b \\ 0 & 1\end{bmatrix}\right)\phi,\psi\rangle$ or $\langle\phi,\psi\rangle = \langle\phi,\rho(\mu_1^{-1},\mu_2^{-1})\left(\begin{bmatrix} 1 & b \\ 0 & 1\end{bmatrix}\right)\psi\rangle$ for all $\phi$ in $\mathcal{H}(\mu_1,\mu_2)$ and $b \in F$.) Since by (4.9) and (4.11) such functions $\psi(g)$ do not exist (unless $\mu(x) = |x|$, in which case $\psi(g)$ must be $\mu_1^{-1}(\det g)|\det g|^{1/2}$) the proof is complete. (Actually there remains the case $\mu(x) = |x|^{-1}$, but by Lemma 4.17 this situation is contragredient to the case $\mu(x) = |x|$ just considered.) □

REMARK 4.19. The principal series representations $\pi(\mu_1,\mu_2)$ and $\pi(\nu_1,\nu_2)$ are equivalent if and only if $(\mu_1,\mu_2) = (\nu_1,\nu_2)$ or $(\nu_2,\nu_1)$. It is also clear from Lemma 4.17 and the proof of Theorem 4.18 that the *special* representations $\pi(\mu_1,\mu_2)$ and $\pi(\mu_2,\mu_1)$ are equivalent.

Given any fixed character $\chi(x)$, it should be noted that there are only finitely many special representations having $\chi$ as central character. (Observe that the central character of any subquotient of $\rho(\mu_1, \mu_2)$ is $\mu_1\mu_2$.)

Now the question remains: *Which irreducible admissible representations of* G *(if any) are not accounted for by the induced representations* $\rho(\mu_1, \mu_2)$ *and their subquotients?*

DEFINITION 4.20. Given an admissible representation $\pi$ of G on V let $V(\pi, N)$ denote the subspace of vectors v in V such that

$$\int_U \pi(n)v\,dn = 0 \tag{4.15}$$

for some open compact subgroup U of N. Then $\pi$ is called *supercuspidal* if $V(\pi, N) = V$.

It is well-known that the matrix coefficients

$$(\pi(g)v, \tilde{v}) \quad v \in V, \quad \tilde{v} \in \tilde{V},$$

of supercuspidal representations of G are compactly supported (modulo the center Z) and that this property characterizes the supercuspidal representations among the irreducible admissible representations of G. Equally well known is the following fact, which is a special case of a general theorem of Jacquet and Harish Chandra.

THEOREM 4.21. *Suppose* $\pi$ *is an irreducible admissible representation of* G. *Then if* $\pi$ *is not supercuspidal it is a subrepresentation of some* $\rho(\mu_1, \mu_2)$ *(i.e. it is equivalent to some principal series or special representation). On the other hand, if* $\pi$ *is supercuspidal, it is not equivalent to a subquotient of any* $\rho(\mu_1, \mu_2)$.

*Proof* (Sketch). Suppose $\pi$ is *not* supercuspidal, i.e. $V(\pi, N)$ is a *proper* subspace of V. Observe then that B leaves $V(\pi, N)$ invariant and N operates trivially on $V/V(\pi,N)$. Therefore $\pi$ induces a representation $\sigma'$ of $B/N(=A)$ in $V/V(\pi,N)$. But by the irreducibility of $\pi$ it follows that $V/V(\pi,N)$ is a B-module of finite type. Thus there is an invariant subspace $V'$ of $V$ which contains $V(\pi,N)$ and is such that the resulting representation $\sigma$ of $B/N$ on $W = V/V'$ is irreducible (and hence one dimensional). Since the linear map

$$v \to f_v(g) = \underline{\pi(g)v} ,$$

where $\underline{v}$ denotes the class of $v$ in $W = V/V'$, is easily seen to intertwine $(\pi, V)$ with the induced representation

$$\operatorname*{Ind}_{B\uparrow G} \sigma ,$$

and since $\sigma$ must be one-dimensional, the proof is complete. □

The question still remains: *Do supercuspidal representations actually exist?* The best way to show that they do is to construct them. As it turns out there are a number of different ways to do this. One involves the very useful notion of "Weil representation." Since this construction is particularly well suited to applications to automorphic forms we present it in Section 7 and again (from a different point of view) in Section 10.

### 3. *Some Properties of Irreducible Admissible Representations*

Here we collect some facts relating to the decomposition of the restriction to K of an irreducible admissible representation of G. Of course one knows *a priori* that any such decomposition contains a given irreducible of K at most finitely many times. But it is a more delicate question to ask *exactly* how many times.

DEFINITION 4.22. An irreducible admissible representation $\pi$ of G is called *class 1* or *spherical* if its restriction to K contains the identity representation at least once.

THEOREM 4.23. *An (infinite-dimensional) irreducible admissible representation* $\pi$ *of* G *is class 1 if and only if* $\pi = \pi(\mu_1, \mu_2)$ *for some pair of unramified characters* $\mu_1, \mu_2$ *of* $F^x$ *and* $\pi$ *is not a special representation. In this case the identity representation is contained exactly once in* $\pi$. *Furthermore, if* $\phi_0(g)$ *denotes any function in the (1-dimensional) subspace of* K*-invariant vectors in* $\mathcal{H}(\mu_1, \mu_2)$, *and* $\tau_\omega$ *denotes the Hecke operator corresponding to convolution over* G *with the characteristic function of the double coset* $K\begin{bmatrix}\varpi & 0\\ 0 & 1\end{bmatrix}K$, *then* $\phi_0(g)$ *is an eigenfunction of* $\tau_\omega$, *and*

$$\phi_0 * \tau_{\varpi}(g) = q^{\frac{1}{2}}\left(q^{s_1} + q^{s_2}\right)\phi_0(g) \tag{4.16}$$

*if* $\mu_i(x) = |x|^{s_i}$. *(Recall that* $|\varpi| = q^{-1}$.)

*Proof.* Most of the statements above are quite well-known and established in a number of places. We content ourselves with the proof of (4.16).

By definition

$$\phi_0 * \tau_{\varpi}(x) = \int_G \phi(xy^{-1})\tau_{\varpi}(y)\,dy\ ,$$

so $\phi_0 * \tau_{\varpi}$ is again obviously right K-invariant. Therefore, assuming the second assertion of the theorem to be true, it follows immediately that $\phi_0$ is an eigenfunction of $\tau_{\varpi}$, thus in particular that $\phi_0 * \tau_{\varpi}(1)$ equals some scalar. But, as in Section 3, we find that

$$K\begin{bmatrix}\varpi & 0\\ 0 & 1\end{bmatrix}K = K\begin{bmatrix}\varpi & 0\\ 0 & 1\end{bmatrix} \cup \bigcup_{\eta \in O_F/P_F} K\begin{bmatrix}1 & \eta\\ 0 & \varpi\end{bmatrix}. \tag{4.17}$$

Therefore, from (4.9), (4.16) and the right K-invariance of $\phi_0$, we have

$$\begin{aligned}\phi_0 * \tau_{\varpi}(1) &= \int_{K\begin{bmatrix}\varpi & 0\\ 0 & 1\end{bmatrix}K} \phi_0(y^{-1})\,dy \\ &= \phi\left(\begin{bmatrix}\varpi^{-1} & 0\\ 0 & 1\end{bmatrix}\right) + \sum_{\eta \in O_F/P_F} \phi\left(\begin{bmatrix}1 & -\eta\\ 0 & \varpi^{-1}\end{bmatrix}\right) \\ &= q^{s_1}q^{\frac{1}{2}} + q\left(q^{s_2}q^{-\frac{1}{2}}\right) = q^{\frac{1}{2}}\left(q^{s_1} + q^{s_2}\right)\end{aligned}$$

as desired. □

Even if $\pi$ is not class 1 we have the following wonderful result.

THEOREM 4.24. *Let $\pi$ denote any irreducible admissible (infinite-dimensional) representation of* G *with central character $\psi$. Then there is a largest ideal $c(\pi)$ of $O_F$ such that the space of vectors* v *with*

$$\pi\left(\begin{bmatrix} a & b \\ c & d \end{bmatrix}\right)v = \psi(a)v \tag{4.18}$$

*for all*

$$\begin{bmatrix} a & b \\ c & d \end{bmatrix} \in \Gamma_0(c(\pi)) = \left\{\begin{bmatrix} a & b \\ c & d \end{bmatrix} \in K : c \equiv 0 (\mathrm{mod}\ c(\pi))\right\} = K^{c(\pi)} \tag{4.19}$$

*is not empty. Furthermore, this space has dimension one.*

REMARK 4.25. This theorem is proved in [Casselman] and is implicit in [Jacquet-Langlands]. Following [Casselman], we call $c(\pi)$ *the conductor of* $\pi$. If $\pi$ is class 1, then $c(\pi)$ is $O_F$. In general, we have the table:

(4.20)

| *Representation* | *Conductor* |
|---|---|
| $\pi = \pi(\mu_1, \mu_2)$ (principal series) | (Conductor of $\mu_1$)(Conductor of $\mu_2$) |
| $\pi = \pi(\mu_1, \mu_2)$ (special representations) | (Conductor of $\mu)^2$ or $\varpi O_F$ |
| $\pi$ supercuspidal | $(\varpi^N O)$, $N \geq 2$ . |

Recall that the conductor of any quasi-character $\mu$ of $F^x$ is the largest ideal $\varpi^n O_F$ such that $\mu$ is trivial on the subgroup $1 + \varpi^n O_F$ of $O_F^x$. For the special representation $\pi(\mu_1, \mu_2)$, the conductor is $\omega O_F$ if and only if $\mu$ is unramified.

As a corollary to the proof of Theorem 4.24 we have that if the ideal $c(\pi)$ is replaced by the ideal $c(\pi)\varpi^i$ contained in it then the dimension of the space satisfying Condition (4.18) for $\begin{bmatrix} a & b \\ c & d \end{bmatrix} \in \Gamma_0(c(\pi)\varpi^i)$ is $(i+1)$.

Theorem 4.24 is crucial to a proper understanding of the correspondence between classical cusp forms and certain representations of GL(2,A).

DEFINITION 4.26. An admissible representation $\pi$ of G on a complex space V will be called *pre-unitary* (or simply *unitary*) if there exists on V an invariant positive-definite hermitian form.

If $\pi$ is pre-unitary, the operators $\pi(g)$ can be extended to unitary operators on the Hilbert space obtained by completing V with respect to this form and the result is a unitary representation in the classical sense. Furthermore, this completion will be topologically irreducible if and only if $(\pi,V)$ is algebraically irreducible.

THEOREM 4.27. *The (pre-)unitary irreducible admissible representations of* G *are:*

(i) *the supercuspidal representations with unitary central character;*

(ii) *the principal series representations* $\pi(\mu_1, \mu_2)$ *with both* $\mu_1$ *and* $\mu_2$ *unitary;*

(iii) *the representations* $\pi(\mu_1, \mu_2)$ *of the principal series for which* $\mu_2(x) = \overline{\mu_1(x)}^{-1}$ *and* $\mu(x) = |x|^{\sigma}$, $0 < \sigma < 1$;

(iv) *the special representations with unitary central character.*

This result is entirely analogous to the real situation (Remark 4.7). Therefore the representations in (ii) and (iii) are called *continuous* (respectively) *complementary* series representations of G. A proof of Theorem 4.26 may be found in [Jacquet-Langlands] or [Godement].

WARNING: For $G = GL(2,F)$ it turns out that every irreducible unitary representation of G is admissible (more precisely, the corresponding representation of G in the complex space of K-finite vectors is admissible in the sense of Definition 4.9). Although it is strongly suspected that any irreducible unitary representation of *any* reductive p-adic group is admissible, at the time of this writing this remains a conjecture (in contrast to the real case).

## C. Global Theory

Suppose $F$ is any *global* field and $v$ is any place of $F$. To put together the local theory just discussed one obviously wants to be able to say that every "nice" representation of $GL(2, A_F)$ is an "infinite tensor product" of certain local representations $\pi_v$ of $G_v = GL(2, F_v)$. If this tensor product is to make sense it should not seem surprising that almost all these $\pi_v$ must be class 1. But rather than continue with these motivating remarks let us simply state some facts.

For each place $v$ of $F$ let $\pi_v$ denote an irreducible unitary (hence admissible) representation of $G_v$ on $H_v$. *We shall assume that the resulting collection* $\{\pi_v\}$ *is such that for almost every* $v$ *the representation* $\pi_v$ *is class* 1. Then one can make sense out of the tensor products $\bigotimes_v H_v$ and $\bigotimes_v \pi_v$ as follows.

Let $S_0$ denote a finite set of places of $F$ containing the archimedean places as well as those where $\pi_v$ is not class 1. For each finite set $S \supset S_0$ set

$$H_S = \bigotimes_{v \in S} H_v \, .$$

For each class 1 $\pi_v$, pick once and for all a unit vector $\xi_v^0$ in the one-dimensional subspace of $K_v$-fixed vectors in $H_v$ (cf. Theorem 4.23). Then for each $S' \supset S$ fix an isometric imbedding of $H_S$ into $H_{S'}$, namely $\xi \to \xi \otimes \bigotimes_{v \in S'-S} \xi_v^0$. This imbedding enables us to define the (pre-) Hilbert space

$$\lim_{\rightarrow} H_S$$

and by completion a Hilbert space

$$H = \hat{\bigotimes_v} H_v \, .$$

The resulting unitary representation

$$\pi = \bigotimes \pi_v \tag{4.21}$$

of $G_A$ in $H$ is given by

$$\pi(g)\xi = \bigotimes_v \pi_v(g_v)\xi_v \tag{4.22}$$

if $\xi = \bigotimes_v \xi_v$ (with $\xi_v = \xi_v^0$ for almost every $v$!) and $g = (g_v)$. This representation is well-defined since $\pi_v(g_v)\xi_v = \xi_v^0$ for almost every $v$ by assumption. It is irreducible because each $\pi_v$ is. Indeed, suppose $(\cdot,\cdot)$ is a continuous hermitian bilinear form on $H$ invariant for $\pi(g)$. Then $(\cdot,\cdot)_S$, its restriction to $H_S$, is invariant for $G_S = \prod_{v \in S} G_v$, and consequently coincides with the natural inner product in $H_S$ (by Schur's Lemma). Since this holds for every finite set $S$, $(\cdot,\cdot)$ must be a multiple of the natural inner product in $H = \lim H_S$. Thus $\pi$ is irreducible.

DEFINITION 4.28. Suppose $\pi$ is an irreducible unitary representation of $G$ on $H$. Let $\sigma$ denote any irreducible representation of $K = \prod_v K_v$ and $H(\sigma)$ the space of vectors in $H$ which transform under $K$ according to $\sigma$. Then $\pi$ is called *admissible* if $H(\sigma)$ is finite dimensional for every such $\sigma$.

Let us observe now that $\pi = \bigotimes_v \pi_v$ is admissible. This follows from the fact that $\sigma = \bigotimes_v \sigma_v$, with $\sigma_v$ the identity representation of $K_v$ for all but finitely many $v$, and the fact that $H(\sigma) = \bigotimes_v H_v(\sigma_v)$.

DEFINITION 4.29. An arbitrary irreducible unitary representation $\pi$ will be called *factorizable* if there are irreducible admissible unitary representations $\pi_v$ of $G_v$ such that $\pi = \bigotimes_v \pi_v$ in the sense of (4.21).

THEOREM 4.30. *Every admissible irreducible unitary representation* $\pi$ *of* $G_A$ *is factorizable, and the corresponding local components* $\pi_v$ *are uniquely determined by* $\pi$.

REMARK 4.31. Theorem 4.30 was first proved by [Gelfand-Graev-Pyatetskii-Shapiro]. A more general form of it, valid for (not necessarily unitarizable) irreducible admissible representations of a certain group algebra of $G_A$, is established in [Jacquet-Langlands]. The reason we work with *unitary* representations now is that (for us) a (generalized) cusp form will be an irreducible constituent of the *unitary* representation provided by the natural action of $G_A$ in the Hilbert space $L^2(G_F \backslash G_A)$. The usefulness of Theorem 4.30 comes from the fact that any such constituent is admissible and hence can be analyzed completely in terms of its local components. (Theorem 4.30 should be viewed as the analogue for GL(2) of the decomposition $\psi = \Pi\psi_v$ valid for any unitary character of $A^x = GL(1, A_F)$.)

FURTHER NOTES AND REFERENCES

*A*. Our treatment of the representation theory of $GL(2, R)$ follows [Jacquet-Langlands] who in turn essentially follow [Bargmann] and [Harish-Chandra 2] (modulo the modifications required by the fact that $GL(2, R)$ is not *quite* the direct product of its center and $SL(2, R)$). In particular, the fact that every admissible representation of $\mathcal{H}(G)$ is a subquotient of some $\rho(\mu_1, \mu_2)$ follows immediately from Harish-Chandra's famous "Subquotient Theorem" specialized to $SL(2, R)$.

*B*. The representation theory of $GL(2, F)$ *over local fields* was initiated by [Mautner] and subsequently developed by several mathematicians (cf. [Shalika], [Tanaka], [Gelfand et al.], [Sally-Shalika], [Silberger], and [Jacquet-Langlands]. Important results for GL(n) appear in [Howe].

The notion of admissibility is implicit in [Mautner] and later works but was first made completely precise (and first *fully* exploited) in [Jacquet-Langlands].

In [Jacquet-Langlands] the classification of admissible representations is accomplished through the study of Kirillov models. Our treatment (by contrast) meticulously avoids Kirillov models (except for the coincidental fact that the space $K(\mu_1, \mu_2)$ which appears in the proof of

Theorem 4.18 provides *the* Kirillov model for the principal series representation $\rho(\mu_1, \mu_2)$!). Our treatment rests instead on the Subquotient Theorem 4.21 (valid more generally for any reductive p-adic group; cf. [Harish-Chandra 3] and [Jacquet]). The drawback of this elementary approach is that Kirillov models do play an indispensable role in the theory of automorphic forms and therefore such models have to be discussed sooner or later. (See Section 6.)

## §5. CUSP FORMS AND REPRESENTATIONS OF THE ADELE GROUP OF GL(2)

Let $R(g)$ denote the natural representation of $GL(2, \mathbb{A})$ in the Hilbert space $L^2(G_{\mathbb{Q}} \backslash G_{\mathbb{A}})$. This representation decomposes as a direct integral

$$(5.1) \qquad \int L^2(G_{\mathbb{Q}} \backslash G_{\mathbb{A}}, \psi) d\psi$$

according to the characters of the center of $G_{\mathbb{A}}$ trivial on $\mathbb{Q}^x$. The resulting representation on each $L^2(G_{\mathbb{Q}} \backslash G_{\mathbb{A}}, \psi)$ will be denoted by $R^{\psi}(g)$.

In the language of representation theory one of the fundamental problems of the modern theory of automorphic forms is the following:

PROBLEM: Decompose $R^{\psi}(g)$ explicitly as a direct sum and integral of irreducible representations.

The rest of these Notes will be devoted in one form or another to a discussion of what is known concerning this decomposition.

It will turn out that the *continuous spectrum* of $R^{\psi}(g)$ is completely known, thanks to the theory of Eisenstein series developed by Selberg and Langlands. The *discrete spectrum*, on the other hand, is not. What is known is that this part of the spectrum is essentially exhausted by the space of cusp forms $L_0^2(G_{\mathbb{Q}} \backslash G_{\mathbb{A}}, \psi)$. Therefore, as in Section 2, it is natural to consider the restriction of $R^{\psi}(g)$ to $L_0^2(G_{\mathbb{Q}} \backslash G_{\mathbb{A}}, \psi)$ and to denote this restriction by $R_0^{\psi}(g)$.

In Section 3 it was found that the classical cusp forms define special elements of $L_0^2(G_{\mathbb{Q}} \backslash G_{\mathbb{A}}, \psi)$. The purpose of the present section is to explain how these forms (and even the real analytic wave forms of Maass) correspond to irreducible constituents of $R_0^{\psi}$ in a natural and precise way.

Consequently classical cusp forms are nothing more nor less than special representations of $GL(2, \mathbb{A})$.

We shall also explain how the decomposition of $R_0^{\psi}(g)$ relates to the decomposition of the natural representation of $SL(2, \mathbb{R})$ in $L_0^2(\Gamma_0(N)\backslash SL(2, \mathbb{R}))$. (See "Notes and References.")

## A. Preliminary Results on the Decomposition of $R_0^{\psi}(g)$.

As already suggested, the main result here is the following adelic version of Part (ii) of Theorem 2.6.

THEOREM 5.1. *The representation* $R_0^{\psi}(g)$ *is the discrete direct sum of irreducible unitary representations each occurring with finite multiplicity.*

This result is entirely functional analytic and a direct consequence of the spectral theory for compact self-adjoint operators. More precisely, its proof follows from the following two well-known lemmas.

LEMMA 5.2. *Suppose* $f$ *is any continuous compactly supported function on* $G_{\mathbb{A}}$. *Then the operator*

$$R_0^{\psi}(f) = \int_{G_{\mathbb{A}}} f(g) R_0^{\psi}(g)\, dg \tag{5.2}$$

*is compact as an operator on* $L_0^2(G_{\mathbb{Q}}\backslash G_{\mathbb{A}}, \psi)$.

*Proof* (Sketch). By definition,

$$R_0^{\psi}(f)\phi(x) = \int_{G_{\mathbb{A}}} f(x^{-1}g)\phi(g)\, dg$$

for each $\phi \in L_0^2(G_{\mathbb{Q}}\backslash G_{\mathbb{A}}, \psi)$. Therefore

$$R_0^{\psi}(f)\phi(x) = \int_{N_{\mathbb{Q}}\backslash G_{\mathbb{A}}} \phi(g) \sum_{N_{\mathbb{Q}}} f(x^{-1}ng)\, dg$$

$$= \int_{N_{\mathbb{Q}}\backslash G_{\mathbb{A}}} \phi(g) K(f, x, g)\, dg$$

where

$$K(f,x,g) = \sum_{n \in N_Q} f(x^{-1}ng) - \int_{N_A} f(x^{-1}ng)\,dn \tag{5.3}$$

*since the cuspidal condition for* $\phi$ *implies that*

$$\int_{N_Q \backslash G_A} \phi(g) \int_{N_A} f(x^{-1}ng)\,dn = 0 .$$

Now for each $c > 0$, let $S(c)$ denote the *Siegel domain* $\omega A_\infty^+(c)K$, where $\omega$ is a relatively compact subset of $N_A A_A$, and $A_\infty^+(c)$ is the set of matrices $\begin{bmatrix} a_1 & 0 \\ 0 & a_2 \end{bmatrix}$ in $A_\infty$ such that $\left|\frac{a_1}{a_2}\right| \geq c$. This domain may be chosen so that $G_A = G_Q S(c)$ and so that $\gamma S(c)$ intersects $S(c)$ for only finitely many $\gamma \in G_Q$.

The burden of the proof of this lemma is to establish an estimate for the kernel $K(f,x,g)$ which implies that for any $x = u\begin{bmatrix} a_1 & 0 \\ 0 & a_2 \end{bmatrix}k$ in $S(c)$, and $\phi$ in $L_0^2(G_Q \backslash G_A, \psi)$,

$$|R_0^\psi(f)\phi(x)| \leq C_N \left|\frac{a_1}{a_2}\right|^{-N} \|\phi\|_2 \tag{5.4}$$

for every $N$. A complete proof of (5.4) may be found in [Langlands] or [Godement 2]. From (5.4) it follows that the set

$$\{R_0^\psi(f)\phi : \|\phi\|_2 \leq 1 \text{ in } L_0^2(G_Q \backslash G_A, \psi)\}$$

is pre-compact. Thus the Lemma follows. □

LEMMA 5.3. *Suppose* $U(g)$ *is a unitary representation of a locally compact group* $G$ *on a Hilbert space* $H$ *with the property that the convolution operator*

$$U(f) = \int_G f(g)\,U(g)\,dg$$

*is compact for every continuous compactly supported function* f *on* G. *Then* H *is the direct sum of countably many invariant subspaces which occur with finite multiplicity.*

The proof of this well-known Lemma may be found in [Gelfand-Graev-Pyatetskii-Shapiro] or [Langlands]. Note that if $f(g) \geq 0$, and $f(g^{-1}) = f(g)$ then $\pi(f)$ is a *self-adjoint* compact operator and consequently has countable discrete spectrum and eigenvalues of finite multiplicity.

REMARK 5.4. Theorem 5.1 is more or less a representation-theoretic statement of the fact that any "classical" space of cusp forms is finite dimensional. (By a "classical" space of forms we understand a subspace of functions in $L^2(\Gamma\backslash SL(2,\mathbb{R}))$ satisfying a K-and $\mathfrak{z}$-finiteness condition.)

A refinement of Theorem 5.1 is provided by the "multiplicity one" result of Jacquet and Langlands (Theorem 5.7 below). This result asserts that any given irreducible unitary representation of $GL(2, \mathbb{A})$ occurs *at most once* in $R_0^{\psi}$. To describe its proof we need to argue as follows.

Let $\mathcal{H}(G_{\mathbb{A}})$ denote the *Hecke group algebra* of $G_{\mathbb{A}}$ defined in the obvious way as the restricted tensor product

$$\bigotimes_p \mathcal{H}(G_p)$$

of the local Hecke group algebras $\mathcal{H}(G_p)$. (Restricted with respect to the unit element $1_p$ of $\mathcal{H}(G_p)$, the characteristic function of the compact subgroup $K_p$.)

Suppose we are given, for each p, an irreducible admissible representation $\pi'_p$ of $\mathcal{H}(G_p)$ on a vector space $V_p$ which for almost every p contains the identity representation of $K_p$. Then we can define an irreducible representation

$$\pi' = \otimes \pi'_p$$

of $\mathcal{H}(G_{\mathbf{A}})$ on $V = \otimes V_p$ in complete analogy with the construction of Section 4.C.

In particular, suppose $\pi$ is an irreducible representation of $G_{\mathbf{A}}$ in the subspace H of $L_0^2(G_{\mathbf{Q}}\backslash G_{\mathbf{A}},\psi)$ and $\otimes\pi_p$ is the factorization of $\pi$. For each p, put

$$V_p = H_p^0 = \underset{\substack{\sigma_p \\ \left(\text{algebraic sum}\right)}}{\bigoplus} H_p(\sigma_p)$$

if $\pi_p$ acts on $H_p$. (The summation extends over all irreducible admissible representations $\sigma_p$ of $K_p$.) Let $\pi'_p$ denote the corresponding representation of $\mathcal{H}(G_p)$ on $V_p$. Then define a representation $\pi' = \otimes\pi'_p$ of $\mathcal{H}(G_{\mathbf{A}})$ on $V = \otimes V_p$ as above. Clearly the isomorphism of H with $\otimes H_p$ (which depended on a choice of $K_p$-fixed vector $\xi_p^0$ for almost every p, cf. Section 4.C) induces a natural isomorphism between V and the space $H^0$ consisting of K-finite functions in H. Thus we have:

LEMMA 5.5. *Every irreducible constituent $\pi$ of the unitary representation $R_0^{\psi}(g)$ determines an (algebraically) irreducible representation $\pi'$ of the group algebra $\mathcal{H}(G_{\mathbf{A}})$ on a space of K-finite functions in $L_0^2(G_{\mathbf{Q}}\backslash G_{\mathbf{A}},\psi)$.*

LEMMA 5.6. *Suppose H is an irreducible subspace of $L_0^2(G_{\mathbf{Q}}\backslash G_{\mathbf{A}},\psi)$. Then functions in the dense subspace $H^0$ (consisting of K-finite vectors) are $C^\infty$ (as functions of the infinite component) and rapidly decreasing in the sense that they satisfy the inequalities*

$$|\phi(x)| \le C_N(\phi)\left|\frac{a_1}{a_2}\right|^{-N} \tag{5.5}$$

*for* $x = u\begin{bmatrix} a_1 & 0 \\ 0 & a_2 \end{bmatrix}k$ *in the Siegel domain* S(c).

*Proof.* Since $\phi$ is K-finite, we may assume that $\phi$ belongs to some subspace $H(\sigma)$ of H consisting of functions which transform under K according to the irreducible representation $\sigma$. But then from the finite dimensionality of $H(\sigma)$ (remember admissibility!) it follows that

$$\phi = R_0^{\psi}(f)\phi$$

for some nice compactly supported function f on $G_A$ (which we may assume $C^\infty$). Therefore the first assertion of the Lemma is obvious and the second is a consequence of the estimate (5.4).

THEOREM 5.7. *The multiplicity of an irreducible constituent* $\pi$ *of* $R_0^{\psi}$ *is one.*

*Proof* (Sketch). Suppose H is the space of $\pi$ in $L_0^2(G_Q \backslash G_A, \psi)$. Let $\pi'$ be the corresponding representation of $\mathcal{H}(G_A)$ on $H^0$ constructed in Lemma 5.5. The functions $\phi$ in $H^0$ being nice by Lemma 5.6, they have Fourier expansions

$$\phi\left(\begin{bmatrix}1 & x\\ 0 & 1\end{bmatrix} g\right) = \sum_{\xi} \phi_\xi(g)\tau(\xi x) \tag{5.6}$$

by Remark 3.5. But $\phi_\xi(g) = 0$ if $\xi = 0$, since $\pi$ occurs in the space of cusp forms.

On the other hand, if $\xi \neq 0$, it follows immediately from the left invariance of $\phi$ under $G_Q$, and the invariance of dx under $x \to \xi x$, that

$$\begin{aligned}
\phi_\xi(g) &= \int_{Q\backslash A} \phi\left(\begin{bmatrix}1 & x\\ 0 & 1\end{bmatrix} g\right) \overline{\tau(\xi x)}\,dx \\
&= \int \phi\left(\begin{bmatrix}\xi & \xi x\\ 0 & 1\end{bmatrix} g\right) \overline{\tau(\xi x)}\,dx \\
&= \int \phi\left(\begin{bmatrix}\xi & x\\ 0 & 1\end{bmatrix} g\right) \overline{\tau(x)}\,dx \\
&= \int \phi\left(\begin{bmatrix}1 & x\\ 0 & 1\end{bmatrix}\begin{bmatrix}\xi & 0\\ 0 & 1\end{bmatrix} g\right) \overline{\tau(x)}\,dx
\end{aligned}$$

therefore

$$\phi\left(\begin{bmatrix}1 & x\\ 0 & 1\end{bmatrix} g\right) = \sum_{\xi \neq 0} W_\phi\left[\begin{pmatrix}\xi & 0\\ 0 & 1\end{pmatrix} g\right] \tag{5.7}$$

where $W_\phi(g)$ is the K-finite $C^\infty$ function defined on $G_{\mathbf{A}}$ by

$$W_\phi(g) = \phi_1(g) = \int_{\mathbf{Q}\backslash\mathbf{A}} \phi\left(\begin{bmatrix}1 & x\\ 0 & 1\end{bmatrix} g\right) \overline{\tau(x)}\, dx . \tag{5.8}$$

Now clearly the correspondence $\phi \to W_\phi$ maps $H^0$ one-to-one onto a space of functions $W(g)$ on $G_{\mathbf{A}}$ satisfying the following properties:

(5.9)

(i) $W\left(\begin{bmatrix}1 & x\\ 0 & 1\end{bmatrix} g\right) = \tau(x) W(g)$ for all $x \in \mathbf{A}$;

(ii) $W\left(\begin{bmatrix}x & 0\\ 0 & x\end{bmatrix} g\right) = \psi(x) W(g)$ for all $x \in \mathbf{A}^x$;

(iii) $W$ is right K-finite, $C^\infty$ as a function of the archimedean component of $G_{\mathbf{A}}$, and rapidly decreasing in the sense of (5.5); and

(iv) The space of these $W(g)$ is preserved by right convolution by elements of $\mathcal{H}(G_{\mathbf{A}})$ and the resulting representation of $\mathcal{H}(G_{\mathbf{A}})$ on this space is equivalent to $\pi'$.

The proof of Theorem 5.7 then rests on the assertion that, *given an irreducible representation* $\pi'$, *the space of functions* $W(g)$ *satisfying (i)-(iv) of (5.9) is unique.* This space of functions is called *the Whittaker space (or Whittaker model) of* $\pi'$. The fact that it is unique could be established now by "elementary" means. However, we prefer to postpone the proof until Section 6 where there is further motivation for considering such models.

In any event (modulo this uniqueness result) the proof of Theorem 5.7 is complete. Indeed if $\pi$ were to occur twice in $R_0^\psi$ we would have two "Whittaker models" for $\pi'$, a contradiction. □

REMARK 5.8. Implicit throughout this discussion (and, in particular, the proof of Lemma 5.5) is the fact that *any irreducible constituent of the unitary representation* $R_0^\psi$ *is admissible, hence factorizable.* In [Gelfand-

Graev-Pyatetskii-Shapiro] it seems to have been proved directly that *every* irreducible unitary representation of $GL(2, \Lambda)$ is factorizable (hence automatically admissible!).

## B. Cusp Forms and Hecke Operators Revisited

In Section 1 we described some special results of Hecke and Atkin-Lehner giving necessary and sufficient conditions for a cusp form to be completely determined by *almost all* its Fourier coefficients. These results may be generalized to cusp forms in $S_k(N, \psi)$ *with arbitrary character* $\psi$ and in this generality may be reformulated and strengthened as follows.

Let $S_k^-(N, \psi)$ denote the subspace of $S_k(N, \psi)$ generated by functions of the form $f(dz)$ where $f(z)$ belongs to $S_k(m, \psi)$ with $m$ a *proper* divisor of $N$ and $d$ any divisor of $N/m$. (This subspace is the zero subspace when $N = 1$, or more generally, when $\psi$ is primitive modulo N.) Let $S_k^+(N, \psi)$ denote the orthocomplement of $S_k^-(N, \psi)$ with respect to the Petersson inner product in $S_k(N, \psi)$. (This is the space of *new forms* in $S_k(N, \psi)$.)

Finally let $L$ denote the algebra generated by the Hecke operators $T(p)$ and $T(p)^*$ where $T(p)^*$ denotes the adjoint of $T(p)$ in $S_k(N, \psi)$ with respect to the Petersson metric. (Recall that $T(p)$ is necessarily normal only if $(p, N) = 1$.)

THEOREM 5.9.

(a) $S_k^+(N, \psi)$ *is the largest* $L$ *invariant subspace of* $S_k(N, \psi)$ *on which* $L$ *is commutative and semi-simple;*

(b) *Let* $L^+$ *denote the restriction of* $L$ *to* $S_k^+(N, \psi)$. *Then the* $\mathbf{C}$*-rank of* $L^+$ *equals the* $\mathbf{C}$*-dimension of* $S_k^+(N, \psi)$;

(c) *Suppose that* $f_1 \in S_k^+(N, \psi)$, $f_2 \in S_k^+(M, \psi)$, $T(p) f_i = a_p^i f_i$, $i = 1, 2$, *for all* $p$,

$$a_p^1 = a_p^2$$

*for almost all* $p$. *Then* $N = M$, *and* $f_1$ *and* $f_2$ *coincide up to a constant multiple.*

The proof of Part (c) involves a comparison of the Dirichlet series attached to $f_1$ and $f_2$ and represents a great triumph of Hecke's theory.

There is a natural *adelic* generalization of Theorem 5.9 which includes the case of the real analytic wave forms of Maass and which we shall presently describe.

Fix $\lambda$ an eigenvalue for the Casimir operator $\Delta$. (Thus $\lambda = -\frac{k}{2}\left(\frac{k}{2}-1\right)$ or $\frac{1-s^2}{4}$, with $s$ purely imaginary or purely real and between $-1$ and $1$.) For each irreducible representation $\sigma$ of $SO(2) \subset K_\infty$ let $H(\psi,\lambda,N,\sigma)$ denote the subspace of functions $\phi$ in $L_0^2(G_Q \backslash G_A, \psi)$ such that

$$\Delta\phi = \lambda\phi \tag{5.10}$$

and

$$\phi(gr(\theta)k_0) = \sigma(r(\theta))\psi(k_0)\phi(g) \tag{5.11}$$

for all $g \in G_A$, $r(\theta) \in SO(2,R)$, and $k_0 \in K_0^N$. (Here $K_0^N$ and $\psi$ are defined as in Section 3.A.) This subspace consists of *cusp forms of type* $(\psi,\lambda,N,\sigma)$.

EXAMPLE 5.10. Suppose $\psi$ corresponds to a character of $(Z/NZ)^x$, $\lambda = -\frac{k}{2}\left(\frac{k}{2}-1\right)$, and $\sigma(r(\theta)) = e^{-ik\theta}$. Then $H(\psi,\lambda,N,\sigma)$ is isomorphic to the classical space $S_k(N,\psi)$ via the map $f \to \phi_f$ defined by (3.4).

EXAMPLE 5.11. Suppose $\psi$ is trivial, $\lambda = \frac{1-s^2}{4}$, and $\sigma$ is the trivial character of $SO(2)$. Then $H(\psi,\lambda,N,\sigma)$ is isomorphic to the space $W_s(\Gamma_0(N))$ consisting of real-analytic cusp forms of level $N$. (The correspondence is simply $f(z) \leftrightarrow \phi(g) = f(g_\infty(i))$.)

Clearly $H(\psi,\lambda,N,\sigma)$ provides a meaningful generalization of the classical notion of cusp form. The natural analogue of the Petersson inner product for $H(\psi,\lambda,N,\sigma)$ is simply

$$(\phi_1,\phi_2) = \int_{Z_A G_Q \backslash G_A} \phi_1(g)\overline{\phi_2(g)}\,dg \tag{5.12}$$

and equipped with this inner product $H(\psi,\lambda,N,\sigma)$ is a finite dimensional Hilbert space.

Note that the integral in (5.12) reduces to the classical integral in (1.11) in case the hypotheses of Example 5.10 are satisfied. Indeed in this case $\phi_1\overline{\phi_2}$ is a function on

$$Z_A G_Q \backslash G_A / K_\infty K_0^N \cong \Gamma_0'(N)\backslash SL(2,R)/SO(2,R) \tag{5.13}$$

and $\phi_1\overline{\phi_2}(g) = f_1(z)\overline{f_2(z)}\,y^k$.

To state the analogue of Theorem 5.9 for the spaces $H(\psi,\lambda,N,\sigma)$ we need first to introduce some Hecke operators appropriate for these spaces.

For each prime $p$, define $\tilde{T}(p)$ and $\tilde{T}(p)^*$ on $H(\psi,\lambda,\sigma,N)$ by

$$(\tilde{T}(p)\phi)(g) = \int_{K_p} \psi^{-1}(k_p)\phi\left(gk_p\begin{bmatrix} p & 0\\ 0 & 1\end{bmatrix}\right)dk_p \tag{5.14}$$

and

$$\tilde{T}(p)^*\phi(g) = \int_{K_p} \psi^{-1}(k_p)\phi\left(gk_p\begin{bmatrix} p^{-1} & 0\\ 0 & 1\end{bmatrix}\right)dk_p \tag{5.15}$$

where $K_p = G_p \cap K_0^N$. These operators are adjoint to one another with respect to the inner product defined by (5.12). If $\psi$ is trivial, $\tilde{T}(p)$ coincides with the Hecke operator $\tilde{T}(p)$ already introduced in Section 3 (cf. (3.15)).

Finally, let $H^-(\psi,\lambda,N,\sigma)$ denote the subspace of $H(\psi,\lambda,N,\sigma)$ generated by $H(\psi,\lambda,m,\sigma)$ for all proper divisors $m$ of $N$ and by elements of $H(\psi,\lambda,N,\sigma)$ which can be expressed as right transforms (by elements of $G_A$) of functions in $H(\psi,\lambda,m,\sigma)$. (Caution: $H(\psi,\lambda,m,\sigma)$ is definitely *not* right $G_A$-invariant!) Let $H^+(\psi,\lambda,N,\sigma)$ denote the orthocomplement of $H^-(\psi,\lambda,N,\sigma)$ in $H(\psi,\lambda,N,\sigma)$ with respect to the inner product (5.12) and let $L$ denote the algebra over $C$ generated by the operators $\tilde{T}(p)$ and $\tilde{T}(p)^*$ acting on $H(\psi,\lambda,N,\sigma)$.

THEOREM 5.12.

(a) $H^+(\psi,\lambda,N,\sigma)$ *is the largest* L*-invariant subspace of* $H(\psi,\lambda,N,\sigma)$ *on which* L *is commutative and semi-simple.*

(b) *If* $L^+$ *denotes the restriction of* L *to* $H^+(\psi,\lambda,N,\sigma)$ *then*

$$\mathrm{rank}_{\mathbb{C}}(L^+) = \dim_{\mathbb{C}} H^+(\psi,\lambda,N,\sigma) .$$

(c) *If* $f_i \in H(\psi,\lambda,N_i,\sigma)$, *and*

$$\tilde{T}(p) f_i = a_p^i f_i \qquad (i = 1, 2)$$

*for all* p, *then* $f_1$ *is a constant multiple of* $f_2$ *as soon as* $a_p^1 = a_p^2$ *for almost all* p.

REMARK 5.13. The maximality assertion of Theorem 5.9(b) (and 5.12(b)) has the following significance. If a cusp form is a eigenfunction for all $\tilde{T}(p)$ and $\tilde{T}(p)^*$, then it is *necessarily* a new form.

Theorem 5.12 is proved in [Miyake] more or less following the classical arguments of Hecke and Atkin-Lehner. The Corollary to this theorem which is of great interest to us and which represents a useful addition to the preliminary results on the decomposition of $R_0^\psi(g)$ is the following:

THEOREM 5.14. *Suppose* $\pi^1$ *and* $\pi^2$ *are two irreducible (unitary) constituents of* $R_0^\psi$. *Suppose also that the local components* $\pi_p^1$ *and* $\pi_p^2$ *are equivalent for all but finitely many finite places. Then* $\pi^1 = \pi^2$.

*Proof.* Certainly it suffices to show that the spaces of $\pi_1$ and $\pi_2$ contain some function common to them both.

So suppose $\sigma$ is an irreducible constituent of the restriction of $\pi_\infty^1 = \pi_\infty^2$ to SO(2). Define the *conductor of* $\pi^i$ to be the integer

$$c(\pi^i) = \prod_{p<\infty} c(\pi_p^i) \tag{5.16}$$

where $c(\pi_p^i)$ is the conductor of $\pi_p^i$ in the sense of Remark 4.25. Then let $H(\psi,\pi^i,c(\pi^i),\sigma)$ denote the subspace of functions in the space of $\pi^i$ in $L_0^2(G_Q \backslash G_A,\psi)$ which satisfy

$$\phi(gr(\theta)k_0) = \sigma(r(\theta))\psi(k_0)\phi(g)$$

for all $g \in G_A$, $r(\theta) \in SO(2,R)$, and $k_0 \in K_0^{N_i}$ with $N_i = c(\pi^i)$. Since $\sigma$ occurs just once in $\pi_\infty^i$, it follows from Theorem 4.24 that

$$\dim H(\psi,\pi^i,c(\pi^i),\sigma) = 1$$

for $i = 1$ and 2. Consequently if $f_i$ spans $H(\psi,\pi^i,N_i,\sigma)$, then for some choice of scalars $\{a_p^i\}$,

$$\tilde{T}(p)f_i = a_p^i f_i$$

*for all* p.

Now suppose the eigenvalue of the Casimir operator corresponding to $\pi_\infty^1 = \pi_\infty^2$ is $\lambda$ (cf. Lemma 2.9). Then clearly $H(\psi,\pi^i,N_i,\sigma) = H(\psi,\lambda,N_i,\sigma)$. Furthermore,

$$a_p^1 = a_p^2$$

*for almost every* p, by our hypothesis on $\pi_p^i$, the fact that almost every $\pi_p^i$ is class 1, and the identity (4.16) which asserts that $a_p^i$ is completely determined by $\pi_p^i$ when $\pi_p^i$ is class 1. Therefore, by Part (c) of Theorem 5.12, $f_1$ is some multiple of $f_2$, and the proof of 5.14 is complete. □

REMARK 5.15. A direct representation-theoretic proof of Theorem 5.14 has been given by [Casselman]. This proof depends upon Jacquet-Langlands' treatment of Dirichlet series associated to cusp forms and will be explained in Section 6.

As a consequence of the proof of Theorem 5.14 we can describe a representation-theoretic formulation of the Conjecture of Ramanujan-Petersson.

LEMMA 5.16. *Suppose* $\pi$ *is an irreducible unitary representation of* $G_A$ *satisfying the following properties:*

(1) $\pi = \bigotimes_p \pi_p$ *occurs in* $R_0^{\psi}$ *with* $\psi$ *trivial on* $R_+^{\times}$;

(2) *the conductor of* $\pi$ *(in the sense of (5.16)) is* N;

(3) $\pi_\infty$ *is equivalent to the discrete series representation* $\sigma(\mu_1, \mu_2)$ *with* $\mu_1\mu_2^{-1}(t) = t^{k-1}\,\mathrm{sgn}(t)$ (cf. Remark 4.6) *and* $\pi_p$ *is equivalent to the class 1 representation* $\pi(s_1^p, s_2^p)$ *for all* p *not dividing* N.

*Let* f(z) *be "the" new form in* $S_k(N, \psi)$ *with the property that* $\phi_f(g)$ *generates the (one-dimensional) subspace of functions* $\phi$ *in the space of* $\pi$ *such that*

$$\phi(gr(\theta)k_0) = \sigma_k(r(\theta))\psi(k_0)\phi(g) \tag{5.17}$$

*for all* $r(\theta) \in SO(2)$ *and* $k_0 \in K_0^N$. *THEN (at last)*

$$T(p)f = C_p f$$

*for all* p, *and for* p *not dividing* N,

$$C_p = p^{\frac{k-1}{2}}\left(p^{s_1} + p^{s_2}\right). \tag{5.18}$$

*Proof.* The Lemma is an immediate consequence of Theorem 4.23 (in particular (4.16)) and Lemma 3.7. The one-dimensionality of the space of functions in H satisfying (5.17) is a consequence of Theorem 4.24 as explained in the proof of Theorem 5.14. □

Recall that T(p) is hermitian whenever $(p,N) = 1$ and $\psi$ is trivial. Therefore each "eigenvalue" $p^{\frac{k-1}{2}}(p^{s_1} + p^{s_2})$ should be real. But $\psi$ trivial implies $s_1 + s_2 = 0$. Consequently $p^{s_1}$ and $p^{s_2}$ are complex conjugates of one another when $\pi(s_1, s_2)$ belong to the continuous series.

When $\pi(s_1, s_2)$ belongs to the complementary series, both $p^{s_1}$ and $p^{s_2}$ are real. Therefore, in either case, $(p^{s_1} + p^{s_2})$ is indeed real.

PROPOSITION 5.17. *Suppose* $\pi$ *is an irreducible constituent of* $R_0^\psi$ *of the type described in Lemma 5.16. Then* $\pi_p$ *belongs to the continuous (as opposed to complementary) series of representations of* $G_p$ *for all* $p$ *not dividing* $N$ *if and only if the Ramanujan-Petersson Conjecture is true.*

*Proof.* From (5.18) it is clear that

$$|C_p| \le 2p^{\frac{k-1}{2}} \quad \text{for } (p, N) = 1 \tag{5.19}$$

if and only if

$$|p^{s_1} + p^{s_2}| \le 2 . \tag{5.20}$$

But (5.19) is precisely the R-P Conjecture, and (5.20) obtains iff $s_1$ and $s_2$ are pure imaginary and index a member of the continuous series. (Cf. Theorem 4.27: if $s_1, s_2$ indexes a complementary series representation, then $s_2 = s_1 - \sigma$ and $\mathrm{Re}(s_1) = \sigma/2$ for some $-1 < \sigma < 1$, $\sigma \neq 0$, so $|p^{s_1} + p^{s_2}| = |p^{s_1}(1+p^{-\sigma})| = p^{\sigma/2} + p^{-\sigma/2} > 2$.) □

REMARK 5.18. The Ramanujan-Petersson Conjecture has an obvious generalization to non-holomorphic cusp forms and for this conjecture Proposition 5.17 (as well as Lemma 5.16) is still valid (with $\pi_\infty$ equivalent now to a principal series representation $\pi(\mu_1, \mu_2)$).

## C. Some Explicit Features of the Correspondence Between Cusp Forms and Representations

Implicit in Subsection 5.B. is the fact that each irreducible constituent $\pi$ of $R_0^\psi$ uniquely determines a new form in $S_k(c(\pi), \psi)$. To complete the description of this correspondence between cusp forms in $S_k(N, \psi)$ and constituents of $R_0^\psi$ it is necessary to proceed in the reverse direction as well, namely from (not necessarily new) forms in $S_k(N, \psi)$ to representations $\pi$.

So suppose that $f(z)$ is an arbitrary form in $S_k(N,\psi)$. By Section 1.C. we may assume, without loss of generality, that $f$ is an eigenfunction for every $T(p)$ with $(p,N) = 1$. Say $T(p)f = a_p f$. Then the corresponding function

$$\phi_f(g) = f(g_\infty(i))\, j(g_\infty, i)^{-k}\psi(k_0)$$

in $L_0^2(G_Q \backslash G_A, \psi)$ is an eigenfunction for $\tilde{T}(p)$ for each $(p,N) = 1$.

Now consider the subspace $H(f)$ of $L_0^2(G_Q \backslash G_A, \psi)$ spanned by translates of $\phi_f(g)$ under $G_A$. Let $\pi_f$ denote the resulting representation of $G_A$ in $H(f)$. Then $\pi_f$ is a direct sum of irreducible unitary representations of $G_A$ each of which occurs in $R_0^\psi$ and *our claim is that each of these irreducible representations has the same local component* $\pi_p$ *for* $p = \infty$ *and all finite* $p$ *not dividing* $N$. From this it follows (by Theorem 5.14) that $\pi_f$ is the direct sum of mutually equivalent constituents of $R_0^\psi$ and hence (by "multiplicity one") that $\pi_f$ itself is an *irreducible* constituent of $R_0^\psi$.

To prove our claim we note first every right translate of $\phi_f(g)$ shares the same eigenvalue for "the" Casimir operator of $G_\infty$. Therefore the infinite component of each summand of $\pi_f$ is completely determined. More precisely, this component must be $\sigma(p,t)$ with $p = k-1$ and $t = 0$. (Compare the notation of Remark 4.6.)

Next we note that the space of each summand of $\pi_f$ contains a function which is right $K_p^N$ invariant and an eigenfunction of $\tilde{T}(p)$ (with eigenvalue $a_p$) for $(p,N) = 1$. The p-component of each summand of $\pi_f$ is then the unique class 1 representation $\pi_p(\mu_1,\mu_2)$ of $G_p$ satisfying $\mu_1\mu_2 = \psi_p$ and $p^{\frac{k-1}{2}}(p^{s_1} + p^{s_2}) = a_p$ if $\mu_i(t) = |t|^{s_i}$. Thus our claim that the p-components of these summands are identical for almost every *finite* $p$ is immediate.

The results above, together with those of Subsection B, can be summarized as follows:

THEOREM 5.19.

(a) *Suppose* f *in* $S_k(N,\psi)$ *is an eigenfunction of* T(p) *for every* p *not dividing* N. *Define* $\pi_f$ *to be the representation of* $G_A$ *determined by the translates of* $\phi_f(g)$ *(cf. 3.4). Then* $\pi_f$ *is irreducible.*

(b) *Conversely, suppose* $\pi$ *is an irreducible constituent of* $R_0^\psi$ *with* $\psi$ *trivial on* $(R_+)^x$. *Define* $\{f_\pi\}$ *to be the equivalence class of the functions* $f_\pi$ *in* $S_k(N,\psi)$ *which share the same eigenvalue for every* T(p) *with* p *relatively prime to* N, *the conductor of* $\pi$, *and which are such that* $\phi_{f_\pi}(g)$ *is contained in the space of* $\pi$. *Then the resulting correspondence*

$$\{f\} \leftrightarrow \pi_f ,$$

*between equivalence classes of cusp forms in the above sense, and irreducible constituents of* $R_0^\psi$, *is one-to-one.*

(c) *The mapping* $f \to \pi_f$, *from functions in* $S_k(N,\psi)$ *to representations of* $G_A$, *is one-to-one when restricted to new-forms in* $S_k(N,\psi)$; *it is not one-to-one when restricted to elements of any old-class in* $S_k(N,\psi)$. *(Cf. Section 1.D. for this terminology.)*

REMARK 5.20. Theorem 5.19 describes the nature of the correspondence between cusp forms and representations. In particular it describes the precise sense in which this correspondence is (and is not) one-to-one.

Since the representation associated to a given form is *uniquely* determined by that form it is natural to ask if there is a recipe for describing the components of this representation in terms of the "parameters" of this form (its weight, level, character, and eigenvalues)? Unfortunately, the answer to this question, in general, seems to be no. However, in certain *special* cases such a recipe *is* possible; some of these cases are described below.

PROPOSITION 5.21. *Suppose* $N = \Pi p_i$ *is a product of distinct primes. If* $f \in S_k(\Gamma_0(N))$ *is a new form and* $T(p)f = a_p f$ *for all* $p$, *then*

$$\pi_f = \bigotimes_p \pi_p$$

*can be described as follows:*

(i) $\pi_\infty$ *is the discrete series representation* $\sigma(p,t)$ *with* $p = k-1$ *and* $t = 0$;

(ii) *if* $(p,N) = 1$, $\pi_p$ *is the class 1 representation* $\pi(\mu_1, \mu_2)$ *such that* $\mu_1\mu_2$ *is trivial and* $a_p = p^{\frac{k-1}{2}}(p^{s_1} + p^{s_2})$ *if* $\mu_i(t) = |t|^{s_i}$; *and*

(iii) *if* $p = p_i$, $\pi_p$ *is "the" special representation of* $G_p$ *with trivial central character.*

*Proof.* Everything is clear from previous discussions except assertion (iii) which follows from the fact that $\phi_f$ is right invariant by

$$K_p^1 = \left\{\begin{bmatrix} a & b \\ c & d \end{bmatrix} \in GL(2, O_p) : c \in pO\right\}.$$

That is, (iii) follows from the fact that $\pi_p$ has conductor 1 (or $pO$). Indeed it is clear from table (4.20) that no supercuspidal representation has conductor 1 and that the principal series representation $\pi(\mu_1, \mu_2)$ cannot have conductor 1 since $\mu_1\mu_2$ must be trivial and therefore $\mu_1$ and $\mu_2$ are either *both* unramified or *both* ramified.

CONCLUDING REMARKS.

(a) The correspondence between cusp forms and representations described in this last section is valid for *non-holomorphic* cusp forms as well. The only modifications required are the obvious ones "at infinity."

(b) *Everything* in this section is valid in the setting of cusp forms defined over *arbitrary* global fields. We have worked exclusively within the context of $\mathbf{Q}$ only because this sometimes simplifies the notation and because it is from this classical context that we shall draw further examples in later sections.

FURTHER NOTES AND REFERENCES

Parts of our discussion in B and C are adapted from [Casselman] and [Miyake].

The relation between the decomposition of $L_0^2(G_Q\backslash G_A,\psi)$ and $L_0^2(\Gamma'_0(N)\backslash SL(2,R))$ can be explained as follows. Suppose $\pi$ is an irreducible constituent of $L_0^2(G_Q\backslash G_A,\psi)$ with conductor $c(\pi)=N$.

Then let $H_\infty$ denote the subspace of the space of $\pi$ consisting of functions in $L_0^2(G_Q\backslash G_A,\psi)$ satisfying

$$\phi(gk_0) = \psi(k_0)\phi(g)$$

for all $k_0 \in K_0^N$. ($K_0^N$ is defined by (4.19).) By Theorem 4.24 this space is equivalent, as a $GL(2,R)$-module, to the representation $\pi_\infty$, the infinite component of $\pi$. But now

$$Z_\infty^+ G_Q\backslash G_A/K_0^N \cong \Gamma'_0(N)\backslash SL(2,R)$$

where $Z_\infty^+ = \left\{\begin{bmatrix} r & 0 \\ 0 & r\end{bmatrix} : r \in (R_+)^x\right\}$. Therefore, supposing only that $\psi$ is trivial on $(R_+)^x$, we conclude that $H_\infty$, *as a* $SL(2,R)$*-module*, is isomorphic to an (essentially) irreducible subspace of $L_0^2(\Gamma'_0(N)\backslash SL(2,R))$ which is equivalent to the restriction of $\pi_\infty$ to $SL(2,R)$. We say "essentially" irreducible because the restriction of $\pi_\infty$ to $SL(2,R)$ decomposes into *two* inequivalent pieces if (and only if) $\pi_\infty$ is a discrete series representation. (For details, see, for example [Gelbart] Chapter II, Section 1.) In any event (either piece of) the restriction of $\pi_\infty$ to $SL(2,R)$ is an irreducible constituent of $L_0^2(\Gamma_0(N)\backslash SL(2,R))$.

In [Langlands] and [Godement 2] Theorem 5.1 is proved in the generality of an arbitrary reductive group G. The fact that $L_0^2(G_Q\backslash G_A,\psi)$ essentially exhausts the discrete spectrum follows from the theory of Eisenstein series and is discussed in Section 8 of these Notes. The characterization of those $\pi$ which occur in $L_0^2(G_Q \backslash G_A, \psi)$ constitutes Jacquet-Langlands' reinterpretation of Hecke's theory and is the proper subject matter of Section 6.

Another group-theoretic treatment of the Ramanujan-Petersson Conjecture (Theorem) appears in [Satake].

## §6. HECKE THEORY FOR GL(2)

Let F denote a global field and $\mathbb{A}$ its ring of adeles. If $\psi$ is grossencharacter of F let $L^2(G_F \backslash G_{\mathbb{A}}, \psi)$ denote the Hilbert space of measurable functions $\phi$ on $GL(2,F)\backslash GL(2,\mathbb{A})$ satisfying

(6.1) $$\phi(zg) = \psi(z)\phi(g) \text{ for all } z \in Z_{\mathbb{A}}$$

and

(6.2) $$\int_{Z_{\mathbb{A}} G_F \backslash G_{\mathbb{A}}} |\phi(g)|^2 \, dg < \infty .$$

As in Section 5, let $R^{\psi}(g)$ denote the unitary representation of $G_{\mathbb{A}}$ given in $L^2(G_F \backslash G_{\mathbb{A}}, \psi)$ by right translation.

DEFINITION 6.1. An irreducible unitary representation $\pi$ of $G_{\mathbb{A}}$ will be called an *automorphic form for* GL(2) if it occurs in some $R^{\psi}$. More precisely, suppose

(6.3) $$R^{\psi}(g) = \left(\bigoplus_j \pi^j\right) \oplus \int \pi^s \, ds .$$

Then $\pi$ is an automorphic form if it is equivalent to some $\pi^j$ *or* $\pi^s$. It is a *cusp form* if it is equivalent to some $\pi^j$ (and not 1-dimensional).

Clearly *certain* cusp forms in the above sense correspond to classical cusp forms in $S_k(N,\psi)$ and these forms have associated to them Dirichlet series which by Hecke and Weil essentially characterize them. (Cf. Sections 1.C and 1.D.)

It might be expected therefore that all generalized cusp forms have associated to them certain Dirichlet series and that these Dirichlet series characterize them in the following sense. Suppose an "L-function" can be attached to a given irreducible unitary representation $\pi$ of $G_A$ and suppose this function satisfies certain prescribed analytic conditions. Then $\pi$ must be a cusp form. That this is indeed the case is the main assertion of Jacquet-Langlands' theory.

The purpose of this section is to describe Jacquet-Langlands' characterization of cusp forms and to explain some implications of their theory for the classical theory of forms.

## A. Hecke Theory for GL(1)

This subsection and the next are included primarily to help motivate the group theoretic techniques of Jacquet and Langlands.

*In this subsection only,* $G_A$ will denote the idele group $A^x = GL(1,A)$ of a number of field $F$. By analogy with Definition 6.1, we shall understand an *automorphic form on* GL(1) to be any irreducible unitary representation of $G_A (= GL(1,A) = A^x)$ which occurs in the decomposition of the natural representation $R(g)$ of $G_A$ in $L^2(G_F \backslash G_A) (= L^2(F^x \backslash A^x)$. But

$$R(g) = \oplus \int_{G_F \backslash G_A} \psi(g)\, d\psi \ . \tag{6.4}$$

Therefore automorphic forms on GL(1) are simply grossencharacters of $F$, i.e. unitary characters of $A^x = G_A$ trivial on $G_F = F^x$, and Hecke's theory of L-functions with grossencharacter is simply a theory of Dirichlet series attached to automorphic forms for GL(1)!

This theory may be described as follows. Suppose $\psi$ is any grossencharacter of $F$. Then there are characters $\psi_v$ of $(F_v)^x$ (for each place $v$ of $F$) such that

$$\psi = \prod_v \psi_v \tag{6.5}$$

with $\psi_v$ unramified for almost every $v$.

Let B denote the (finite) set of places of F "ramified" for $\psi$, i.e. such that $\psi_v$ is ramified. For each v *not* in B, put

$$\chi(v) = \psi_v(\varpi_v) \tag{6.6}$$

where $\varpi_v$ denotes a local uniformizing parameter at v (a generator of the prime ideal of $O_v$). Note that $\chi(v)$ does *not* depend on the choice of $\varpi$. Indeed $\psi_v$ is trivial on the units of $F_v$. Therefore $\chi$ is well defined by multiplicativity on the group of F-ideals relatively prime to B (the "bad" primes) and the character $\chi$ is the *Hecke character* associated to the grossencharacter $\psi$.

The *Hecke* L-*series associated to* $\chi$ (or $\psi$) is defined by the Dirichlet series

$$L(s,\chi) = \sum \frac{\chi(\mathfrak{A})}{N(\mathfrak{A})^s}, \tag{6.7}$$

where the summation is taken over all F-ideals relatively prime to B, or, equivalently, by the Euler product

$$L(s,\chi) = \prod_{v \notin B} \left(1 - \frac{\chi(v)}{(N(v))^s}\right)^{-1} = \prod_{v \notin B} \left(1 - \frac{\psi_v(\omega_v)}{|\omega_v|^{-s}}\right)^{-1}. \tag{6.8}$$

Although it is easy to check that this series (or product) converges absolutely uniformly to the right of the line $\mathrm{Re}(s) = 1$ it is not so easy to prove Hecke's principal result concerning this series.

THEOREM 6.2.

(i) $L(s,\chi)$ *has an analytic continuation as a meromorphic function of* s. *If* $\chi$ *is not "trivial" then* $L(s,\chi)$ *is actually entire; otherwise* $L(s,\chi)$ *has a simple pole at* $s = 1$ *as its only singularity;*

(ii) $L(s,\chi)$ *satisfies a functional equation of the following type: there is a constant* A, *and a gamma factor* $\Gamma(s,\chi)$ *(depending only on the infinite primes of* F) *such that if*

$$R(s,\chi) = s(s-1)A^s\,\Gamma(s,\chi)L(s,\chi)$$

*then* $R(s,\chi)$ *is entire and*

$$R(1-s,\chi^{-1}) = W(\chi)R(s,\chi) \tag{6.9}$$

*with* $W(\chi)$ *a constant of modulus* 1.

Hecke's proof of this result uses the classical theory of theta functions and represents a great achievement.

The arithmetic significance of Theorem 6.2 is well-known. Indeed if $\chi$ is trivial, $L(s,\chi)$ is the Dedekind zeta-function of F. Thus, taking $F = \mathbb{Q}$, Theorem 2 gives (in particular) the analytic continuation and functional equation of the Riemann zeta function.

A drawback of Hecke's proof is that it does not lay bare the group theoretic interpretation of the result. This was left to be done by Tate in his justly celebrated thesis. Since Tate's method has since proved to be germinal to many developments in number theory (let alone Jacquet-Langlands theory) it seems worthwhile to sketch it below.

Let $S(\mathbb{A})$ denote the *Schwartz-Bruhat* functions on $\mathbb{A}$. These functions are linear combinations of functions of the form

$$\prod_v f_v$$

where

(i) $f_v$ is in the usual Schwartz space of $\mathbb{R}$ or $\mathbb{C}$ if v is archimedean;

(ii) $f_v$ belongs to $S(F_v)$, the space of locally constant compactly supported functions on $F_v$ if v is finite; and

(iii) $f_v$ is the characteristic function of the ring of integers of $F_v$ for almost every finite v.

If $\tau$ is any fixed non-trivial character of $\mathbb{A}$ trivial on F then the Fourier transform of any function f in $S(\mathbb{A})$ is given by

$$\hat{f}(\alpha) = \int_{\mathbb{A}} f(\beta)\tau(\alpha\beta)\,d\beta\ . \tag{6.10}$$

Now suppose $\psi$ is an automorphic form on $G_A$, i.e. $\psi$ is a grossencharacter of F. For each f in S(A) *the global zeta-function* $\zeta(f,\psi,s)$ is defined by

$$(6.11) \qquad \zeta(f,\psi,s) = \int_{G_A = A^\times} f(a)\psi(a)|a|^s d^\times a$$

where $|a| = \prod_v |a_v|_v$ and $d^\times a = \prod_v d^\times a_v$. If $\psi = \prod_v \psi_v$ and $f = \prod_v f_v$ it is not difficult to show that *for* $\mathrm{Re}(s) > 1$,

$$(6.12) \qquad \prod_v \int_{F_v} |f_v(a_v)|a_v|_v^s| d^\times a_v < \infty$$

and

$$(6.13) \qquad \zeta(f,\psi,s) = \prod_v \zeta(f_v,\psi_v,s) ,$$

where $\zeta(f_v,\psi_v,s)$ denotes the *local zeta-function*

$$(6.14) \qquad \zeta(f_v,\psi_v,s) = \int_{(F_v)^\times = G_v} f_v(a_v)\psi_v(a_v)|a_v|_v^s d^\times a_v .$$

Therefore $\zeta(f,\psi,s)$ is easily seen to represent an analytic function of s *for* $\mathrm{Re}(s) > 1$. What Tate proved was:

THEOREM 6.3. *The function* $\zeta(f,\psi,s)$ *has an analytic continuation to the whole* s*-plane, as an entire function if* $\psi$ *is non-trivial, and as a meromorphic function with simple poles at* 0 *and* 1 *otherwise. Furthermore,* $\zeta(f,\psi,s)$ *satisfies the functional equation*

$$(6.15) \qquad \zeta(f,\psi,s) = \zeta(\hat{f},\psi^{-1},1-s) .$$

COROLLARY 6.4. *The function* $\pi^{-s/2}\Gamma\left(\frac{s}{2}\right)\zeta(s)$ *extends to a meromorphic function in* $\mathbb{C}$ *(with poles at* 0 *and* 1*) which is invariant under the change* $s \to 1-s$.

*Proof*. Take $F = \mathbb{Q}$, $\psi$ trivial, and $f = \prod_p f_p$, where $f_\infty(x) = e^{-\pi x^2}$, and $f_p$ is the characteristic function of $O_p$. Then (choosing $\tau$ as in Section 3)

$$\zeta(f_\infty, \psi_\infty, s) = \pi^{-s/2}\Gamma\left(\frac{s}{2}\right)$$

and

$$\zeta(f_p, \psi_p, s) = \int_{O_p} |\alpha|^s \, d^x\alpha = (1-p^{-s})^{-1}$$

Consequently

$$\zeta(f, \psi, s) = \pi^{-s/2}\Gamma\left(\frac{s}{2}\right)\zeta(s)$$

and the Corollary follows immediately from Theorem 6.4. □

COROLLARY 6.5. *Theorem 6.2 is valid.*

*Proof* (Sketch). Using harmonic analysis on $G_v = (F_v)^x$ it can be shown that the *local* zeta-functions themselves possess analytic continuations to the $\mathbb{C}$-plane and satisfy functional equations of the form

$$\zeta(f_v, \psi_v, s) = \rho(\psi_v, s)\zeta(\hat{f}, \psi_v^{-1}, 1-s) \tag{6.16}$$

where $\rho(\psi_v, s)$ is a meromorphic function of $s$ *which does not depend on* $f_v$. (The "additive" Fourier transform $\hat{f}_v$ is defined with respect to the character $\tau_v$ if $\tau = \prod \tau_v$.)

A straightforward computation then shows that for *properly chosen* $f_v$

$$\prod_{v \notin B_\psi} \zeta(f_v, \psi_v, s) = L(s, \chi) \prod_{v \notin B_\psi} \chi(\partial_v)^{-1} N(\partial_v)^{s-\frac{1}{2}} \tag{6.17}$$

where $B_\psi$ is the union of the set of archimedean primes and finite primes ramified for $\psi$ and $\partial_v$ is the different of $F_v$. But

$$\zeta(f,\psi,s) = \prod_v \zeta(f_v,\psi_v,s) .$$

Therefore, by multiplying both sides of (6.17) by $\prod_{v \in B_\psi} \zeta(f_v,\psi_v,s)$, it follows that $L(s,\chi)$ (like $\zeta(f,\psi,s)$ *and* $\prod_{v \in B_\psi} \zeta(f_v,\psi_v,s)$) possesses an analytic continuation to all of $\mathbb{C}$. Similarly the functional equation for $L(s,\chi)$ results from the functional equation for the global *and* local zeta-functions $\zeta(f,\psi,s)$. (The *precise* analytic behavior of $L(s,\chi)$ as well as an effective computation of $A$, $\Gamma(s,\chi)$ and $W(\chi)$ follows from a careful analysis of the factors $\rho(\psi_v,s)$.) □

REMARK 6.6. (Concerning the proof of Theorem 6.3.) A complete proof of Theorem 6.3 is to be found in several places (see "Notes and References"). We wish to emphasize that the proof is almost entirely Fourier analytic. More precisely, it represents a blend of (additive and multiplicative) harmonic analysis on the groups $M = M(1,F)$ and $G = GL(1)$, the main ingredients being the Poisson summation formula, some generalized Mellin transforms, and a computation of the measure of the homogeneous space

$$Z_\infty^+ G_F \backslash G_{\mathbb{A}} = R_+ F^\times \backslash \mathbb{A}^\times .$$

We should also remark that nowhere in the proof of Theorem 6.3 do the local zeta-functions $\zeta(f_v,c_v,s)$ play a real role. In fact these zeta-functions first significantly appear in the proof of Corollary 6.5. Thus in Tate's work the local zeta-functions are principally introduced to show that the theory is non-empty!

CONCLUDING REMARKS 6.7. It seems to make sense to ask if there is a "Converse" to Hecke's theory for $GL(1)$. In other words, suppose $\psi$ is a character of $G_{\mathbb{A}} = \mathbb{A}^\times$ such that $L(s,\chi)$ (defined as in 6.7) satisfies the analytic properties announced in Theorem 6.2. Is $\psi$ necessarily a grossencharacter, i.e. is $\chi$ necessarily trivial on $F^\times$?

## B. Further Motivation

It is clearly indicated now that one should associate to *every* automorphic form $\pi$ an L-function with functional equation. Indeed this is what Hecke did for GL(1) *and* GL(2) *in case* $\pi$ corresponds to a classical holomorphic cusp form. It is also indicated that one should characterize the L-functions so obtained *for general* $\pi$ following Weil's Theorem 1.12. Both these goals are attained by Jacquet-Langlands' theory. They work in the generality of an arbitrary global field and their new methods involve the theory of group representations.

In Subsection C we shall sketch some principal results of Jacquet-Langlands' theory. Since their methods seem complex at first sight we shall attempt in the present subsection to provide motivation for them.

How does Hecke's theory attach a Dirichlet series to f in $S_k(N,\psi)$? According to (1.24), one simply takes the Mellin transform of f along the line $\{iy : y > 0\}$. That is,

$$L(s,f) = \int_0^\infty f(iy)\, y^{s-1}\, dy \ . \tag{6.18}$$

(Strictly speaking, the *Dirichlet series* attached to f is $D(s,f) = (2\pi)^s\,\Gamma(s)^{-1}L(s,f)$. However our interest is in the *L-function* $L(s,f)$, a simple modification of $D(s,f)$.)

To understand how one might generalize Hecke's construction it is helpful to rewrite (6.18) in the framework of the adele group of GL(2).

As in Section 3, let

$$\phi_f(g) = f(g_\infty(i))\, j(g_\infty, i)^{-k}\psi(k_0) \tag{6.19}$$

denote the function on $G_A = GL(2,A)$ corresponding to f in $S_k(N,\psi)$. For simplicity, suppose that $\psi \equiv 1$ and $N = 1$. Then $\phi_f(g)$ is right invariant by $K_0 = \prod_{p<\infty} K_p$ (and, as always, left invariant by $G_Q$). From (6.19) it follows that (for $y > 0$)

$$\phi_f\left(\begin{bmatrix} y & 0 \\ 0 & 1 \end{bmatrix}\right) = f(iy) \ .$$

Therefore (6.18) is equivalent to the formula

$$L(s,f) = \int_{Q^\times \backslash A^\times} \phi\left(\begin{bmatrix} y & 0 \\ 0 & 1 \end{bmatrix}\right) |y|^s d^\times y \tag{6.20}$$

(Recall that $A^\times = Q^\times R_+ (\prod_{p<\infty} O_p^\times)$.)

To continue "adelizing" (6.18) one must exploit the existence of a Fourier expansion for $f(z)$. According to Lemma 3.6,

$$\begin{aligned} \phi\left(\begin{bmatrix} y & 0 \\ 0 & 1 \end{bmatrix}\right) &= \sum_{\xi \in Q^\times} \phi_\xi\left(\begin{bmatrix} y & 0 \\ 0 & 1 \end{bmatrix}\right) \\ &= \sum_{n=1}^{\infty} a_n e^{-2\pi n y} = f(iy) \end{aligned} \tag{6.21}$$

Equivalently,

$$\phi\left(\begin{bmatrix} y & 0 \\ 0 & 1 \end{bmatrix}\right) = \sum_{\xi \in Q^\times} W_\phi\left(\begin{bmatrix} \xi y & 0 \\ 0 & 1 \end{bmatrix}\right) \tag{6.22}$$

where $W_\phi(g)$ is the K-finite $C^\infty$ function defined on $G_A$ by

$$W_\phi(g) = \int \phi\left(\begin{bmatrix} 1 & x \\ 0 & 1 \end{bmatrix} g\right) \overline{\tau(x)} dx = \phi_1(g) . \tag{6.23}$$

Consequently, from (6.20),

$$L(s,f) = \int_{Q^\times \backslash A^\times} \sum_{\xi \in Q^\times} W_\phi\left(\begin{bmatrix} y\xi & 0 \\ 0 & 1 \end{bmatrix}\right) |y|^s dy^* ,$$

or

$$L(s,f) = \int_{A^\times} W_\phi\left(\begin{bmatrix} y & 0 \\ 0 & 1 \end{bmatrix}\right) |y|^s d^* y . \tag{6.24}$$

What (6.24) says is that $L(s,f)$ is the adelic Mellin transform "along $\begin{bmatrix} y & 0 \\ 0 & 1 \end{bmatrix}$" of the first Fourier coefficient of a special function $\phi_f$ in the space of $\pi_f$.

Suppose now that $\pi$ is an *arbitrary* constituent of some $R_0^{\psi}$. From the classical considerations above it is suggested that one *might* define $L(s,\pi)$ as the "Mellin transform" of the first Fourier coefficient of some appropriate $\phi(g)$ in the space of $\pi$. *But the obvious problem is to choose this* $\phi(g)$ *or* $W(g)$ *correctly.*

To gain some insight into this problem let us examine the special properties satisfied by $W_\phi(g)$ in case $\phi = \phi_f$. As in Section 5.B, it is easy to verify that the space of right translates of $W_\phi(g)$ generates a space of functions $W(\pi_f)$ on $G_{\mathbf{A}}$ satisfying the following properties:

(i) $W\left(\begin{bmatrix} 1 & x \\ 0 & 1 \end{bmatrix} g\right) = \tau(x)W(g)$ for all $x \in \mathbf{A}$;

(ii) $W\left(\begin{bmatrix} z & 0 \\ 0 & z \end{bmatrix} g\right) = \psi(z)W(g)$ for all $z \in \mathbf{A}^\times$;

(iii) $W$ is right K-finite, $C^\infty$ as a function of $G_\infty$, and rapidly decreasing; and

(iv) the natural representation of the Hecke algebra $\mathcal{H}(G_{\mathbf{A}})$ on $W(\pi_f)$ given by right convolution is equivalent to $\pi_f$.

(Strictly speaking, $W(\pi_f)$ should be defined as the space of right *convolutions* of $W_\phi$ with elements of $\mathcal{H}(G_{\mathbf{A}})$ since the action of $G_{\mathbf{A}}$ at the infinite places does *not* preserve $K_\infty$-finiteness.)

This realization of the representation $\pi_f$ is called the *Whittaker model* for $\pi_f$ and $W(\pi_f)$ is called its *Whittaker space*. Thus $L(s,f)$ may be viewed described as the Mellin transform (along $\begin{bmatrix} y & 0 \\ 0 & 1 \end{bmatrix}$) of a certain privileged function in the Whittaker space of $\pi_f$.

What is special about $W_{\phi_f}(g)$ (among other things) is that it is right $K_0$ invariant. Thus, given an arbitrary constituent $\pi$ of $R_0^{\psi}$, one presumably could compute the conductor $c(\pi)$ of $\pi$ and select a "correct" $K_0^{c(\pi)}$ invariant function $W(g)$ in the Whittaker space of $\pi$. Then one could introduce $L(\pi,s)$ as some kind of Mellin transform of $W(g)$. This is in fact what Jacquet-Langlands do except that their attack on the problem gets closer to the heart of the matter. Jacquet-Langlands characterize special functions $W_v(g_v)$ in the Whittaker space of each local component $\pi_v$ of $\pi$ and then piece together these local functions to get $W(g)$; these local functions in turn are characterized in terms of the local "zeta functions" described in Subsection C below.

### C. Jacquet-Langlands' Theory

Suppose $\pi$ is any irreducible unitary (admissible) representation of $G_A = GL(2, A)$. The purpose of this subsection is to explain how Jacquet-Langlands attach to each such $\pi$ an L-function $L(s, \pi)$ and how they characterize those $\pi$ that occur in $R_0^\psi$ in terms of the functions $L(s,\pi)$.

Analogous to Hecke's theory for $GL(1)$ the first step will be to exploit the factorization $\pi = \otimes \pi_v$ announced in Theorem 4.30. (Cf. (6.5).) It should not seem surprising therefore that a "global" L-function $L(\pi,s)$ is defined by first introducing "local L-factors" $L(\pi_v, s)$ for each $\pi_v$ and then piecing these factors together. (Cf. the definition (6.8).) As already suggested, these local factors are described by introducing Whittaker models for each $\pi_v$ and then computing the Mellin transform of a certain privileged function in this space.

To make matters precise, fix once and for all a non-trivial additive character $\tau$ of $F\backslash A$ and suppose $\tau(x) = \Pi \tau_v(x_v)$ for $x = (x_v)$ in $A$.

THEOREM 6.8 (Existence and Uniqueness of the Local Whittaker Model). *Suppose* $\pi_v$ *is an irreducible admissible (infinite-dimensional) representation of* $G_v, v$ *any place of* $F$. *Then in the space of locally constant functions on* $G_v$ *such that*

$$W\left(\begin{bmatrix} 1 & x \\ 0 & 1 \end{bmatrix} g\right) = \tau_v(x) W(g) \tag{6.25}$$

$(x \in F_v, g \in G_v)$ *there is a unique subspace* $W(\pi_v)$ *which is invariant for the right action of* $G_v$ *and equivalent (as a* $G_v$*-module) to* $\pi_v$.

*As in the global case,* $W(\pi_v)$ *is called the Whittaker space of* $\pi_v$, *and* $(G_v, W(\pi_v))$ *its Whittaker model.*

REMARK 6.9. In case $v$ is archimedean some obvious modifications and additions are required. In this case we should consider irreducible admissible representations *of the group algebra* $\mathcal{H}(G_v)$ and consider only solutions of (6.25) which are $C^\infty$ and such that

(6.26) $$W\left(\begin{bmatrix} t & 0 \\ 0 & 1 \end{bmatrix}\right) = O(|t|)^N \text{ as } |t| \to \infty .$$

Such solutions actually define classical Whittaker functions (along $\begin{bmatrix} t & 0 \\ 0 & 1 \end{bmatrix}$). (Hence the terminology for arbitrary fields.) The correct statement of Theorem 6.8 *for archimedean places* is that there is a unique space of such functions on which the natural action of $\mathcal{H}(G_v)$ is equivalent to $\pi_v$.

REMARK 6.10 (*Concerning the proof of Theorem 6.8*). Jacquet-Langlands' proof of Theorem 6.8 is simplest to explain in the archimedean case. In this case every $\pi_v$ is a subrepresentation of some $\rho(\mu_1, \mu_2)$ on $\mathcal{H}(\mu_1, \mu_2)$ (in the notation of Section 4). Thus (as in the proof of Theorem 4.18) one can introduce the space of functions

(6.27) $$W_\phi(g) = \int \phi\left[w^{-1}\begin{pmatrix} 1 & x \\ 0 & 1 \end{pmatrix} g\right] \overline{\tau_v(x)}\, dx$$

with $\phi$ in the subspace of $\mathcal{H}(\mu_1, \mu_2)$ belonging to $\pi_v$. This space of functions can be shown to provide a Whittaker model for $\pi_v$. To establish uniqueness one uses the differential equations on p. 4.5 to show that such a space of functions has a basis each of whose elements satisfy a certain second order differential equation (coming from the Casimir operator pushed down to the symmetric space $G/K$) with only one solution (of sufficiently small growth at $\infty$! Cf. (6.26)).

In the non-archimedean case a variant of the method of proof just sketched is required.

The significance of Theorem 6.8 for the theory of automorphic forms has already been suggested and should now be apparent. Indeed the local *uniqueness* of Whittaker models implies global uniqueness. Hence the multiplicity one result of Section 5 follows immediately. On the other hand local *existence* also makes possible the choice of a priviledged function in the space of $\pi_v$ and hence ultimately the choice of a function in the space of $\pi$ whose Mellin transform will coincide with $L(s,\pi)$. To explain this in more detail we need first to introduce the local and global "zeta functions" for $\pi_v$ and $\pi$.

DEFINITION 6.11. Let $\pi_v$ denote an irreducible admissible representation of $G_v$ and $W(\pi_v)$ its Whittaker space. Suppose $\chi$ is a unitary character of $(F_v)^x$, $g \in G_v$, $W \in W(\pi_v)$, and $s$ a complex number. Then the *local zeta function* attached to $(g,\chi,W)$ is defined by the formula

$$\zeta(g,\chi,W,s) = \int_{F_v^x} W\left(\begin{bmatrix} a & 0 \\ 0 & 1 \end{bmatrix} g\right) \chi(a) |a|^{s-\frac{1}{2}} d^x a \,. \tag{6.28}$$

CAUTION: Although $\xi(a) = W\left(\begin{bmatrix} a & 0 \\ 0 & 1 \end{bmatrix} g\right)$ is a relatively well behaved function on $F^x$ it is not necessarily locally constant on $F$. Therefore the function of $s$ defined by (6.28) does *not* coincide with the local zeta-function (6.14) introduced for $GL(1)$!

THEOREM 6.12.

(i) *The integral defining* $\zeta(g,\chi,W,s)$ *converges for* $s$ *with sufficiently large real part;*

(ii) *There exists an Euler factor* $L(s,\chi\otimes\pi_v)$ *with the property that*

$$\frac{\zeta(g,\chi,W,s)}{L(s,\chi\otimes\pi_v)}$$

*is an entire function of* $s$ *for every* $g$, $\chi$, *and* $W$, *and such that*

$$\zeta(1,\chi,W^0,s) = L(s,\chi\otimes\pi_v) \tag{6.29}$$

*for an appropriate choice of* $W^0 \in W(\pi_v)$;

(iii) *The function* $\zeta(g,\chi,W,s)$ *possesses an analytic continuation to the whole s-plane and satisfies the functional equation*

$$\frac{\zeta(g,\chi,W,s)}{L(s,\chi\otimes\pi_v)}\,\varepsilon(\pi_v,\chi,s) = \frac{\zeta(wg,\chi^{-1}\psi_v^{-1},W,1-s)}{L(1-s,\psi_v^{-1}\chi_v^{-1}\otimes\pi_v)}$$

*where* $\varepsilon(\pi_v,\chi,s)$ *is independent of* $g$ *and* $W$, $\psi_v$ *is the central character of* $\pi_v$, *and* $w = \begin{bmatrix} 0 & +1 \\ -1 & 0 \end{bmatrix}$.

REMARK 6.13 (*Concerning the statement of Theorem 6.12*).

(i) By an *Euler factor* we understand a function of $s$ of the form $P^{-1}(q^s)$ where $P$ is a polynomial such that $P(0) = 1$ and $q = |\tilde{\omega}_v|^{-1}$ if $v$ is finite. For archimedean places $v$ the factor $L(s,\pi)$ will be a product of certain gamma functions to be specified below. In either case $L(s,\pi)$ *is unique* up to a constant factor. Indeed if $L(s,\pi)$ and $L^*(s,\pi)$ are two Euler factors satisfying the conditions of the theorem their quotient is an entire function without zeros.

(ii) The contragredient representation to $\pi_v$ is simply

(6.31) $$\tilde{\pi}(g) = \psi_v^{-1}(g)\pi_v(g)$$

if $\psi_v$ is the central character of $\pi_v$. Therefore (6.30) reads

(6.32) $$\frac{\zeta(W,s)}{L(s,\pi_v)}\,\varepsilon(\pi_v,s) = \frac{\zeta(\psi_v^{-1}W',1-s)}{L(1-s,\tilde{\pi}_v)}$$

if $g = 1$, $\chi = 1$, and $W'$ denotes the right translate of $W$ by $\begin{bmatrix} 0 & 1 \\ -1 & 0 \end{bmatrix}$ (a kind of Fourier transform). The significance of (6.32) is explained by recalling the local theory of Tate alluded to in Subsection A.

Tate's theory asserts that there is an Euler factor $L(s,\psi)$ such that (in the notation of (6.14))

$$\frac{\zeta(f_v,\psi_v,s)}{L(s,\psi)}$$

is entire in $s$ for every $f \in S(F_v)$ and

$$\frac{\zeta(f_v,\psi_v,s)}{L(s,\psi_v)}\,\varepsilon(\psi_v,s) = \frac{\zeta(f_v,\tilde{\psi}_v,1-s)}{L(1-s,\tilde{\psi}_v)} \tag{6.33}$$

for some (exponential) function $\varepsilon(\psi_v,s)$ which is independent of s. (Here $\bar{\psi}_v = \psi_v^{-1}$ = the "contragredient" of the representation $\psi_v$.) In fact some straightforward computation shows that

$$L(s,\psi_v) = (1-\psi_v(\tilde{\omega})q^{-s})^{-1} \tag{6.34}$$

if $\psi_v$ is unramified and

$$L(s,\psi_v) \equiv 1 \tag{6.35}$$

otherwise. In either case

$$L(s,\psi_v) = \zeta(f_v,\psi_v,s) \tag{6.36}$$

for an appropriate choice of $f_v$ and the analogy between this situation and the situation for GL(2) described by Theorem 6.12 is immediate.

For the record,

$$\varepsilon(\psi,s) \equiv 1$$

*if* $\psi$ *is unramified* and

$$\varepsilon(\psi,s) = (\text{root of unity})\,q^{\mathrm{cond}(\psi)(\frac{1}{2}-s)}$$

in general.

(iii) The factor $L(s,\chi\otimes\pi_v)$ may be interpreted as a g.c.d. for the family of meromorphic local zeta-functions $\zeta(g,W,\chi,s)$.

(iv) The special function $W_v^0(g)$ satisfying (6.29) is the privileged local Whittaker function alluded to earlier. It will be the *global* function

$$W^0(g) = \prod_v W_v^0(g) \tag{6.37}$$

whose adelic Mellin transform will coincide with $L(s,\pi)$.

THEOREM 6.14. *Suppose two irreducible admissible representations* $\pi^1$ *and* $\pi^2$ *of* $G_v$ *induce the same central character. Then* $\pi^1$ *and* $\pi^2$ *are equivalent if and only if*

$$\varepsilon(\chi\otimes\pi^1, s)\frac{L(1-s,\chi^{-1}\otimes\tilde{\pi}^1)}{L(s,\chi\otimes\pi^1)} = \varepsilon(\chi\otimes\pi^2, s)\frac{L(1-s,\chi^{-1}\otimes\tilde{\pi}^2)}{L(s,\chi\otimes\pi^2)} \tag{6.38}$$

*for every character* $\chi$ *of* $F_v^x$.

This theorem says that *the local factors* $L(\chi\otimes\pi_v, s)$ *and* $\varepsilon(\chi\otimes\pi_v, s)$ *uniquely determine* $\pi_v$ among all the representations $\pi_v$ with given central character. Actually, given any $\pi_v$, these local factors can be explicitly described as follows.

THEOREM 6.15 (Description of $L(s,\pi)$ and $\varepsilon(s,\pi)$: The Non-Archimedean Case). *In the table below we shall use the notation of Section 4; the factor* $L(s,\psi)$ *is the local factor for* GL(1) *given by (6.34) and (6.35); finally*

$$E(\mu_1,\mu_2,s) = \frac{L(1-s,\mu_1^{-1})}{L(s,\mu_2)} = \begin{cases} 1 & \textit{if } \mu_1 \textit{ is ramified} \\ -\mu_2(\tilde{\omega})q^{-s} & \textit{otherwise} \end{cases}$$

TABLE

| $\pi$ | $L(s,\pi)$ | $\varepsilon(s,\pi)$ |
|---|---|---|
| $\pi = \pi(\mu_1,\mu_2)$ (principal series) | $L(s,\mu_1)L(s,\mu_2)$ | $\varepsilon(s,\mu_1)\varepsilon(s,\mu_2)$ (equals 1 if $\mu_1$ and $\mu_2$ are unramified) |
| $\pi = \pi(\mu_1,\mu_2)$ (special representation) | $L(\mu_1,s)$ | $\varepsilon(s,\mu_1)\varepsilon(s,\mu_2)E(\mu_1,\mu_2,s)$ |
| $\pi$ supercuspidal | 1 | unimportant to us |

THEOREM 6.16 (Description of $L(s,\pi)$ and $\varepsilon(s,\pi)$: The Archimedean Case). *We restrict our attention to the case* $F = \mathbb{R}$:

| $\pi$ | $L(s,\pi)$ | $\varepsilon(s,\pi)$ |
|---|---|---|
| $\pi = \pi(\mu_1,\mu_2)$ (principal series) | $L(s,\mu_1)L(s,\mu_2)$ | $\varepsilon(\mu_1,s)\varepsilon(\mu_2,s)$ |
| $\pi = \sigma(p,t)$ (discrete series representation in $\pi(\mu_1,\mu_2)$) | $(2\pi)^{-s-\frac{(t+p)}{2}}\Gamma\left(s+\frac{t+p}{2}\right)$ | $i^{(p+1-n_1-n_2)}\,i^{n_1+n_2}$ $(\mu_i(x) = \lvert x_i\rvert^{s_i}\,\mathrm{sgn}(x)^{n_i})$ |

Here

(6.39) $$L(s,\mu_j) = \pi^{-\frac{1}{2}(s+r_j+m_j)}\,\Gamma\left(\frac{s+r_j+m_j}{2}\right)$$

and

(6.40) $$\varepsilon(\mu_j,s) = i$$

if $\mu_j(x) = |x|^{r_j}\,\mathrm{sgn}(x)^{m_j}$, $m_j = 0,1$.

These tables will be helpful when we compare Jacquet-Langlands' theory with Hecke's classical theory. It will also be helpful to have the following information concerning the special function $W^0(g)$.

PROPOSITION 6.17 (Description of $W^0(g)$).

(a) *Archimedean case* $F_v = \mathbb{R}$: *If* $\pi_v$ *belongs to the principal series for* $G_v$ *let* $W_v^0(g)$ *denote the unique* $O(2)$*-invariant function in* $W(\pi_v)$; *then* $W_v^0(g)$ *satisfies the relation*

(6.41) $$\zeta(1,\chi,W^0,s) = L(s,\chi\otimes\pi_v)\ .$$

*If* $\pi_v$ *is equivalent to the discrete series representation* $\sigma(p,t)$ *(in the notation of Remark 4.6) then (6.41) is valid with* $W_v^0(g)$ *equal to the unique "lowest weight vector" in* $W(\pi_v)$, *i.e. the function in* $W(\pi_v)$ *such that* $\rho(U)\phi = i(p+1)\phi$ *(cf. the notation of p. 57).*

(b) *Non-archimedean* $F_v$: *If* $\pi_v$ *is class one then* $W(\pi_v)$ *contains exactly one* $K_v$ *invariant function* $W_v^0(g)$ *such that* $W_v^0(e) = 1$ *and for this* $W_v^0(g)$ *(6.29) obtains.*

*If* $\pi_v$ *is not class then (6.29) obtains with* $W_v^0(g)$ *equal to the function in* $W(\pi_v)$ *which is (almost) right-invariant for* $K_p^{c(\pi_v)}$. *See (4.18) and (4.19).*

CAUTION: Parts of the last proposition are simply *implicit* in Chapter I of [Jacquet-Langlands].

We shall now explain how these local results are pieced together to obtain a global theory which generalizes the classical theory of Hecke.

Suppose $\pi$ is an irreducible unitary admissible representation of $GL(2, \mathbb{A})$. By Theorem 4.30,

$$\pi = \otimes \pi_v$$

where almost every $\pi_v$ is class 1. On the other hand, every $\pi_v$ in (6.42) possesses a *unique* Whittaker model $W(\pi_v)$ by Theorem 6.8, and the space of functions of $G_{\mathbb{A}}$ generated by functions of the form

$$W(g) = \prod_v W_v(g_v) ,$$

where $W_v = W_v^0$ for almost every $v$, provides a Whittaker model $W(\pi)$ for $\pi$. In particular the action of $\mathcal{H}(G_{\mathbb{A}})$ on $W(\pi)$ given by right convolution is equivalent to $\pi$. What is important is that this space $W(\pi)$ is unique. Indeed any space of functions on $G_{\mathbb{A}}$ satisfying properties (i)-(ii) of (5.9) must by local uniqueness coincide with $W(\pi)$.

We remind the reader that *if* $\pi$ *occurs in* $R_0^\psi$ *then* an alternate construction of $W(\pi)$ results from the consideration of the first Fourier coefficients of functions in the space of $\pi$. More precisely, suppose $H^0$ denotes the space of K-finite functions in the space of $\pi$. The map

$$\phi(g) \to \phi_1(g) = \int \phi\left(\begin{bmatrix} 1 & x \\ 0 & 1 \end{bmatrix} g\right) \overline{\tau(x)}\, dx$$

then defines a one-to-one correspondence between $\pi$ and $W(\pi)$.

To state the main theorem of Jacquet-Langland's theory we set

$$\varepsilon(\pi,\chi,s) = \prod_v \varepsilon(\pi_v \otimes \chi_v, s) \tag{6.42}$$

for each grossencharacter $\chi = \Pi\chi_v$ of F. Since $\pi_v$ is almost always class 1 this product is actually a finite product by Theorem 6.15. We also put

$$L(s,\chi\otimes\pi) = \prod_v L(s,\chi_v \otimes \pi_v)\ . \tag{6.43}$$

This product converges for Re(s) sufficiently large since for almost every v, $L(s,\chi_v\otimes\pi_v)$ is of the form

$$[1-\mu_v\chi_v(\tilde{\omega}_v)q^{-s}]^{-1}[1-\nu_v\chi_v(\tilde{\omega}_v)q^{-s}]^{-1}$$

where $\mu_v$ and $\nu_v$ are local quasi-characters and $|\mu_v(x)| = |\nu_v(x)| = |x|^{\pm\sigma/2}$, $0 \le \sigma \le 1$.

THEOREM 6.18. *Suppose* $\pi = \otimes\pi_v$ *has central character* $\psi$. *Then* $\pi$ *occurs in* $R_0^\psi$ *if and only if* $L(s,\chi\otimes\pi)$ *satisfies the following properties for every grossencharacter* $\chi$ *of* F:

(i) $L(s,\chi\otimes\pi)$ *extends to an entire function bounded in vertical strips; and*

(ii) $L(s,\chi\otimes\pi)$ *satisfies the functional equation*

$$L(s,\chi\otimes\pi) = \varepsilon(\pi,\chi,s)\,L(s,\chi^{-1}\otimes\tilde{\pi}) \tag{6.44}$$

*where* $\tilde{\pi}(g) = \psi^{-1}(g)\pi(g)$.

*Proof* (Sketch). Suppose first that $\pi$ occurs in $R_0^\psi$ and that its space is H. For each v choose a function $W_v(g_v)$ in $W(\pi_v)$ such that $W_v = W_v^0$ almost everywhere. Then

$$W(g) = \prod W_v(g_v)$$

certainly belongs to $W(\pi)$ and

$$\phi(g) = \sum_{\xi \in F^{\times}} W\left(\begin{bmatrix} \xi & 0 \\ 0 & 1 \end{bmatrix} g\right) \tag{6.45}$$

belongs to the space of K-finite functions in H.

Next consider the Mellin transform

$$\zeta(g, W, \chi, s) = \int_{F^{\times}\backslash A^{\times}} \phi\left(\begin{bmatrix} x & 0 \\ 0 & 1 \end{bmatrix} g\right) \chi(x) |x|^{s-\frac{1}{2}} d^{\times}x \ . \tag{6.46}$$

Since $\phi\left(\begin{bmatrix} x & 0 \\ 0 & 1 \end{bmatrix} g\right)$ is rapidly decreasing at $\infty$ (and at 0 by its right $G_F$-invariance) the integral in (6.46) converges for all values of s and represents an entire function bounded in vertical strips. But $W\left(\begin{bmatrix} x & 0 \\ 0 & 1 \end{bmatrix} g\right)$ also is rapidly decreasing at $\infty$. Therefore

$$\int_{A^{\times}} W\left(\begin{bmatrix} x & 0 \\ 0 & 1 \end{bmatrix} g\right) \chi(x) |x|^{s-\frac{1}{2}} d^{\times}x = \prod_{v} \zeta(g_v, \chi_v, W_v, s)$$

converges for Re(s) sufficiently large and in this range equals $\zeta(g, W, \chi, s)$. (Here $\zeta(g_v, \chi_v, W_v, s)$ is the local zeta function defined by (6.28).)

Now suppose $W_v(g)$ chosen so that

$$\zeta(g_v, \chi_v, W_v, s) = L(s, \chi_v \otimes \pi_v) \ .$$

With this choice of $W(g) = \Pi\, W_v(g_v)$ we have (from above) that for R(s) sufficiently large,

$$\zeta(g, W, \chi, s) = \prod_{v} L(s, \chi_v \otimes \pi_v) = L(s, \chi \otimes \pi) \ .$$

Consequently $L(s, \chi \otimes \pi)$ obviously satisfies Condition (i) of Theorem 6.28 since $\zeta(g, \chi, W, s)$ does. To verify (ii) observe that

$$\zeta(g, \chi, W, s) = \zeta(wg, \psi^{-1}\chi^{-1}, W, 1-s)$$

since $\phi(wg) = \phi(g)$. Consequently by the *local functional equations*,

$$\frac{\zeta(wg, \psi^{-1}\chi^{-1}, W, 1-s)}{L(s, \chi^{-1} \otimes \tilde{\pi})} = \prod_v \frac{\zeta(wg_v, \chi_v^{-1}\psi_v^{-1}, W_v, 1-s)}{L(s, \chi_v^{-1} \otimes \tilde{\pi}_v)}$$

$$= \prod_v \varepsilon(s, \chi_v, \pi_v) \frac{\zeta(g, \chi_v, W_v, s)}{L(s, \chi_v \otimes \pi_v)}$$

$$= \varepsilon(s, \chi, \pi) \frac{\zeta(g, \chi, W, s)}{L(s, \chi \otimes \pi)}$$

$$= \varepsilon(s, \chi, \pi) \frac{\zeta(wg, \psi^{-1}\chi^{-1}, W, 1-s)}{L(s, \chi \otimes \pi)}$$

and from this equality the functional equation (6.44) is immediate.

Conversely, suppose $L(s, \chi \otimes \pi)$ satisfies Conditions (i) and (ii) of Theorem 6.28 for all $\chi$. To prove that $\pi$ then occurs in $R_0^\psi$ we simply "reverse" the procedure just outlined.

For each $W \in W(\pi)$ put

$$\phi_W(g) = \sum_{\xi \in F^\times} W\left(\begin{bmatrix} \xi & 0 \\ 0 & 1 \end{bmatrix} g\right) . \tag{6.47}$$

What has to be shown is that the map

$$W \to \phi_W$$

imbeds $W(\pi)$ into $L_0^2(G_F \backslash G_A, \psi)$ and this reduces ultimately to showing that

$$\phi_W(wg) = \phi_W(g) . \tag{6.48}$$

Indeed the invariance of $\phi_W$ on the left by rational matrices of the form $\begin{bmatrix} a & b \\ 0 & c \end{bmatrix}$ is almost trivial from (6.47) (assuming the convergence of (6.47) has been established). Therefore Bruhat's decomposition together with known properties of $W(g)$ implies that (6.48) is equivalent to $\phi$ belonging to $L_0^2(G_F \backslash G_A, \psi)$.

To prove (6.48) it certainly suffices to show that

$$\phi_W\left(\begin{bmatrix} x & 0 \\ 0 & 1 \end{bmatrix} g\right) = \phi_W\left(w\begin{bmatrix} x & 0 \\ 0 & 1 \end{bmatrix} g\right) \tag{6.49}$$

*for all* $x \in A^x$. To prove (6.49) it suffices in turn to verify that the Mellin transforms of the left and right hand sides are equal. Now the Mellin transform of the left hand side of (6.49) is simply

$$\zeta(g,\chi,W,s) = \int_{A^x} W\left(\begin{bmatrix} x & 0 \\ 0 & 1 \end{bmatrix} g\right) \chi(x)|x|^{s-\frac{1}{2}} d^x x$$

$$= \prod_v \zeta(g_v,\chi_v,W_v,s) \qquad (\text{for } \operatorname{Re}(s) >> 0)$$

$$= L(s,\chi\otimes\pi) \prod_v \frac{\zeta(g_v,\chi_v,W_v,s)}{L(s,\chi_v\otimes\pi_v)},$$

an entire function by virtue of the local theory *and* our assumptions on $L(s,\chi\otimes\pi)$. On the other hand the Mellin transform of the right hand side is

$$\int_{F^x\backslash A^x} \phi_W\left(w\begin{bmatrix} x & 0 \\ 0 & 1 \end{bmatrix} g\right) \chi(x)|x|^{s-\frac{1}{2}} d^x x$$

$$= \int \phi_W\left(\begin{bmatrix} x^{-1} & 0 \\ 0 & 1 \end{bmatrix} wg\right) \psi(x)\chi(x)|x|^{s-\frac{1}{2}} d^x x$$

$$= \int \phi_W\left(\begin{bmatrix} x & 0 \\ 0 & 1 \end{bmatrix} wg\right) \psi^{-1}(x)\chi^{-1}(x)|x|^{\frac{1}{2}-s} d^x x$$

$$= \zeta(wg,\psi^{-1}\chi^{-1},W,1-s)$$

$$= L(s,\chi^{-1}\otimes\tilde{\pi}) \prod_v \frac{\zeta(wg_v,\chi_v^{-1}\psi_v^{-1},W_v,1-s)}{L(s,\chi_v^{-1}\otimes\tilde{\pi}_v)}.$$

Therefore both Mellin transforms are entire functions and the fact that they coincide is an immediate consequence of the local *and* global functional equations. □

EXAMPLE 6.19 (*Holomorphic Cusp Forms*). Suppose $\pi_f = \otimes \pi_p$ is the representation of $GL(2, \mathbb{A})$ generated by the cusp form $f(z)$ in $S_k(SL(2, \mathbb{Z}))$ whose eigenvalues for $T(p)$ are

$$a_p = p^{\frac{k-1}{2}} \left(p^{s_1} + p^{-s_1}\right). \tag{6.50}$$

By Jacquet-Langlands' theory,

$$L(\pi, s) = \prod_{p \leq \infty} L(s, \pi_p)$$

$$= (2\pi)^{-s-\left(\frac{k-1}{2}\right)} \Gamma\left(s + \frac{k-1}{2}\right) \prod_{p < \infty} \left(1-p^{-s_1}p^{-s}\right)^{-1} \left(1-p^{s_1}p^{-s}\right)^{-1}.$$

(Cf. (6.34), Theorem (6.15), and Theorem (6.16).) Therefore by (6.50),

$$L(s, \pi) = (2\pi)^{-s-\left(\frac{k-1}{2}\right)} \Gamma\left(s + \frac{k-1}{2}\right) \prod \left(1-p^{-s-\frac{k-1}{2}} a_p + p^{-2s}\right)^{-1}$$

an L-function *which agrees with Hecke's* L-*function*

$$L(s', f) = (2\pi)^{-s'} \Gamma(s')$$

with $s' = s + \frac{k-1}{2}$. (Cf. (1.28).) Since in this case $\varepsilon(s, \pi) = i^k$, *the functional equation (6.44) is nothing but the functional equation (1.27) of Hecke*. (The substitution $s \to 1-s$ in (6.44) corresponds to the substitution $s' \to k-s'$.) Consequently Jacquet-Langlands' theory is consistent with (at least this part of) Hecke's theory.

EXAMPLE 6.20 (*Maass' Wave-forms*). In [Maass] certain L-functions are attached to real analytic cusp-forms in $W_s(SL(2, \mathbb{Z}))$. As in Example 6.19 it can be shown that these L-functions (and their functional equations) appear as special cases of the general theory sketched above.

CONCLUDING REMARKS. The "real" reason why the L-functions $L(s,\pi_f)$ and $L(s,f)$ of Example 6.19 coincide is the following. For an appropriately chosen $W(g) = \Pi W_v(g)$ in $W(\pi_f)$, namely $W(g) = W_\phi(g)$ with $\phi = \phi_f$,

$$L(s,\pi) = \zeta(1,W,1,s)$$

(6.51)
$$= \int_{\mathbf{A}^\times} W\left(\begin{bmatrix} x & 0 \\ 0 & 1 \end{bmatrix}\right) |x|^{s-\frac{1}{2}} d^\times x .$$

But since $W_v^0(g)$ is right $K_0$ invariant and transforms under $SO(2)$ according to the character $e^{-ik\theta}$,

$$L(s,\pi) = \int_{\mathbf{Q}^\times\backslash\mathbf{A}^\times} \phi\left(\begin{bmatrix} y & 0 \\ 0 & 1 \end{bmatrix}\right) |y|^{s-\frac{1}{2}} d^\times y$$

with

$$\phi(g) = \phi_f(g) = f(g_\infty(i))\, j(g_\infty, i)^{-k} ,$$

or

$$L(s,\pi) = \int_0^\infty f(iy)\, y^{s+\left(\frac{k-1}{2}\right)-1} dy = L(s',f)!$$

The equality (6.51) furthermore emphasizes the fact that in proving the functional equation for $L(s,\pi)$, Jacquet-Langlands directly follow Hecke. Indeed it is (a generalization of) the Mellin transform which (once again) translates the automorphy condition for $\phi$ into a functional equation for $L(s,\pi_\phi)$. In establishing their "Converse to Hecke Theory" Jacquet-Langlands follow Weil.

## D. Connections with the Classical Theory

Here we shall use Jacquet-Langlands' theory to derive at least one "new" result in the classical theory of forms.

The first natural question to ask is whether Theorem 6.18 actually generalizes or simply overlaps Hecke's theory. In particular, does $L(s,\pi_f)$ always agree with Hecke's $L(s', f)$! All we know thus far is that *if* f *is of level* 1 then these functions *do* coincide.

So suppose now that f belongs to $S_k(\Gamma_0(N))$. Then even in the context of the classical theory there is more than one way to attach an L-function to f. On the one hand, we have

$$L(s,f) = \int_0^\infty f(iy)\, y^{s-1}\, dy \ .$$

On the other, there is no loss of generality in assuming that

$$T(p)f = a_p f \tag{6.52}$$

*for all* p *relatively prime to* N. Then there are two possibilities. Either f is a *new form*, in which case (6.52) holds *for all* p, or f is an *old* form, in which case $f(z) = h(dz)$ with h a new form in $S_k(\Gamma_0(M))$ and

$$T(p)h = a_p^* h \qquad \textit{for all } p \ .$$

If the first possibility obtains, set

$$L^*(s,f) = (2\pi)^{-s}\Gamma(s)\prod(1-a_p p^{-s}+p^{k-1-2s})^{-1} \ .$$

If the second does, set

$$L^*(s,f) = (2\pi)^{-s}\Gamma(s)\prod_{p<\infty}(1-a_p^* p^{-s}+p^{k-1-2s})^{-1} .$$

Note that $a_p^* = a_p$ for all $(p,N) = 1$. The natural question to ask is whether $L^*(s,f)$ coincides (up to a constant multiple) with $L(s,f)$ and the obvious answer is "*if and only if* f *is a new form.*"

PROPOSITION 6.21. $L(s,\pi_f)$ *and* $L(s',f)$ *coincide (up to a constant multiple) if and only if* f *is a new form in* $S_k(\Gamma_0(N))$.

*Proof* (Sketch). Suppose f is a new form. Then by Proposition 6.17, and the discussion of Subsection C,

$$L(s,\pi_f) = \int_{\mathbf{Q}^\times\backslash\mathbf{A}^\times} \phi_f\left(\begin{bmatrix} y & 0 \\ 0 & 1\end{bmatrix}\right)|y|^{s-\frac{1}{2}}\,dy$$

$$= \int_0^\infty f(iy)\,y^{s+\left(\frac{k-1}{2}\right)-1}\,dy = L(s',f)\ .$$

On the other hand, suppose f is an old form belonging to the new form h. Then

$$L(s,\pi_f) = L(s,\pi_h) = L(s',h) \neq L(s',f)\ . \square$$

The simple application of Jacquet-Langlands' theory we have in mind is reminiscent of Theorem 5.14 whose representation-theoretic proof was promised earlier.

THEOREM 5.14. *Suppose* $\pi^1$ *and* $\pi^2$ *are two irreducible unitary constituents of* $R_0^\psi$. *Suppose also that the local components* $\pi_v^1$ *and* $\pi_v^2$ *are equivalent for all but finitely many finite places. Then* $\pi^1 = \pi^2$.

*Proof* (Sketch). Let S denote the finite set of finite places where it is not assumed $\pi_v^1 = \pi_v^2$. To prove that $\pi_v^1 = \pi_v^2$ for p in S we shall use the fact that the factors $L(\chi_v \otimes \pi_v, s)$ and $\varepsilon(\chi_v \otimes \pi_v, s)$ completely determine $\pi_v$ (Theorem 6.14).

Because $\pi^1$ and $\pi^2$ occur in $R_0^\psi$ the L-functions $L(S,\chi\otimes\pi^1)$ and $L(s,\chi\otimes\pi^2)$ satisfy functional equations of the type specified by Jacquet-Langlands' Theorem 6.18. Dividing the functional equation for $\pi^1$ by the functional equation for $\pi^2$, and rearranging slightly, we have

$$\prod_{v \in S} \frac{L(1-s,\chi_v^{-1}\otimes\tilde{\pi}_v^1)\varepsilon(s,\chi_v\otimes\pi_v^1)}{L(s,\chi_v\otimes\pi_v^1)}$$

$$= \prod_{v \in S} \frac{L(1-s,\chi_v^{-1}\otimes\tilde{\pi}_v^2)\varepsilon(s,\chi_v\otimes\pi_v^2)}{L(s,\chi_v\otimes\pi_v^2)} .$$

To conclude that $\pi_v^1 = \pi_v^2$ for each v in S one has only to eliminate all terms on each side of (6.53) except the one involving v and then apply Theorem 6.14. To do this one appeals, in turn, to the following two facts from [Jacquet-Langlands]: (1) For any place w of F and any irreducible admissible representation $\pi_w$,

$$L(s,\chi_w\otimes\pi_w) = 1$$

and

$$\varepsilon(s,\chi_w\otimes\pi_w) = \varepsilon(s,\chi_w\psi_w)\varepsilon(s,\chi_w)$$

for $\chi_w$ a quasicharacter of $F_w^x$ *with sufficiently large conductor*; and (2) For any character $\delta$ of $F_w^x$ we can find a grossencharacter of F which restricts to $\delta_u$ on $F_u^x$ and has arbitrarily large conductor at the other places of S. □

The "new" result we have in mind is:

PROPOSITION 6.22. *Suppose $\pi^1$ and $\pi^2$ are two irreducible unitary constituents of $R_0^\psi$. Suppose also that the local components $\pi_p^1$ and $\pi_p^2$ are equivalent for all finite places. Then $\pi_\infty^1$ is equivalent to $\pi_\infty^2$ and $\pi^1 = \pi^2$. (We are working over* **Q**!)

*Proof.* By Theorem 6.18 the L-functions $L(s,\pi^i)$ corresponding to $\pi^i$ satisfy the functional equation

$$\frac{L(1-s,\chi^{-1}\otimes\pi^i)}{L(s,\chi\otimes\pi^i)}\,\varepsilon(s,\chi\otimes\pi^i) = 1$$

for all grossencharacters $\chi = \Pi\chi_p$ of **Q**.

But by assumption, the quotients

$$\frac{L(1-s,\chi_p^{-1}\otimes\pi_p^i)\varepsilon(s,\chi_p\otimes\pi_p^i)}{L(s,\chi_p\otimes\pi_p)} \qquad i=1,2$$

coincide for *all finite* p. Therefore they coincide for $p=\infty$ as well, so by Theorem 6.14, $\pi_\infty^1=\pi_\infty^2$. □

COROLLARY 6.23. *Suppose* $f(z)$ *is a real analytic wave form for* $\Gamma_0'(N)$ *with* N *square free. Then if*

$$T(p)f = a_p f \qquad \textit{for all } p ,$$

*knowledge of* $\{a_p\}$ *completely determines the eigenvalue of* f *for* $\Delta$.

The special case of this corollary for $N=1$ was communicated to me by Cartier before I had stated or proved Proposition 6.22. Its proof in general is similar to the proof of the following refinement of Hecke's Corollary 1.10.

THEOREM 6.24. *The Hecke ring has multiplicity one in the graded ring of holomorphic cusp forms of integral weight. More precisely, if* $f_i \in S_{k_i}(N,\psi)$ *and*

$$T(p)f_i = a_p f_i \qquad \textit{for all } p ,$$

*then* $k_1=k_2$, *and* $f_1$ *is a multiple of* $f_2$.

*Proof.* To give a representation-theoretic proof (i.e. to apply Proposition 6.22) we need to assume N square free. In this case the representation $\pi_f$ corresponding to f is determined at *every finite* place and hence at infinity as well. To prove the Theorem in general one argues directly with the functional equations for the Euler products

$$(2\pi)^{-s}\Gamma(s)\prod_p (1-a_p p^{-s}+p^{k-1-2s})^{-1}$$

*after* making the changes of variables $s\to s'+\frac{k_i-1}{2}$.

FURTHER NOTES

The analogy between Jacquet-Langlands' theory and Tate's proof of Theorem 6.2 should not be overlooked. In fact it is possible to give another proof of the "only if" part of Theorem 6.18 which completely avoids Whittaker models and follows Tate's proof *directly* rather than by analogy. This alternate approach is of considerable interest and has led to a fair amount of recent research which we shall at least sketch in the paragraphs below.

The basic idea is to introduce the *local zeta functions*

$$\zeta(f_v, \pi_v, u, \tilde{u}, s) = \int_{G_v} f_v(x) < \pi_v(x) u, \tilde{u} > |\det(x)|^{s+\frac{1}{2}} dx$$

for each Schwartz-Bruhat function $f_v$ on $M(2, F_v)$, each irreducible admissible representation $(\pi_v, V_v)$ of $G_v$, $u \in V_v$, $\tilde{u} \in \tilde{V}_v$, and $s \in \mathbf{C}$. These integrals can be shown to converge for $\mathrm{Re}(s)$ sufficiently large, to continue meromorphically to the whole s-plane, and to satisfy the functional equation

$$(*) \qquad \frac{\zeta(f_v, \pi_v, u_v, \tilde{u}_v, s)}{L(s, \pi_v)} \varepsilon(s, \pi) = \frac{\zeta(\hat{f}_v, \tilde{\pi}_v, u, \tilde{u}_v, 1-s)}{L(1-s, \tilde{\pi}_v)} .$$

Here $\hat{f}_v$ denotes the (additive) Fourier transform of $f_v$, and $L(s, \pi)$ and $\varepsilon(s, \pi)$ are as in Theorem 6.18. The quotients

$$\frac{\zeta(f_v, \pi_v, u_v, \tilde{u}_v, s)}{L(s, \pi_v)} = \Xi(s, f_v, u_v, \tilde{u}_v)$$

are entire for *all* $f_v$ and

$$\sum_{i=1}^{n} \Xi(s, f_v, u_i, \tilde{u}_i)$$

is actually a non-zero constant for *some* choice of $f_v$, $u_1, \cdots, u_n$, and $\tilde{u}_1, \cdots, \tilde{u}_n$.

The above results extend Tate's local theory for GL(1) to GL(2) (cf. Section 6.A). What is remarkable is that the functions $\zeta(f_v, \pi_v, u, \tilde{u}, s)$ and $\zeta(g, 1, W, s)$ (cf. Definition 6.11) lead to precisely the same Euler factors $L(s, \pi_v)$.

Now if $\pi = \bigotimes_v \pi_v$ is an irreducible admissible representation of $G_{\mathbf{A}}$ which occurs in $L_0^2(G_F \backslash G_{\mathbf{A}}, \psi)$ the global zeta-function

$$\zeta(f, \pi, u, \tilde{u}, s) = \int_{G_{\mathbf{A}}} f(x) < \pi(x) u, \tilde{u} > |\det(x)|^s \, dx$$

is defined for $f \in S(M(2, \mathbf{A}))$, $u$ and $\tilde{u}$ in the dense subspace of $\psi$-cusp-forms, and $s$ in some half-plane. But by the Poisson summation formula, this function is shown to extend to an *entire* function in all of $\mathbf{C}$ which satisfies the functional equation

$$\zeta(f, \pi, u, \tilde{u}, s) = \zeta(\tilde{f}, \tilde{\pi}, u, \tilde{u}, 2-s)$$

(cf. (6.15)). Thus by proceeding just as Tate did for GL(1) one concludes (from the global *and* local theory) that the infinite product

$$L(s, \pi) = \prod_v L(s, \pi_v)$$

satisfies all the properties announced in Theorem 6.18.

This proof of the "only if" part of Theorem 6.18 is to be found in Section 13 of [Jacquet-Langlands]. Its generalization to arbitrary simple algebras is the subject matter of [Godement-Jacquet]. The special case of division quaternion algebras is also described in Section 14 of [Jacquet-Langlands] and Section 10 of these Notes.

Yet another approach to the functional equation (*) is possible for certain *operator-valued* zeta functions attached to *unitary* representations. For the sake of definiteness, suppose that $G = GL(2, \mathbf{R})$ and $\pi$ is an irreducible unitary representation belonging to the discrete series for G.

If $f \in S(M(2,R))$ the function $f(x)\,|\det(x)|^s$ belongs to $L^1(G)$ whenever $\mathrm{Re}(s) > 1$ and therefore the zeta-function integral

$$\zeta(f,\pi,s) = \int_G f(x)\pi(x)\,|\det(x)|^s\,dx$$

defines a holomorphic (operator-valued) function of s in this range. The analytic continuation and functional equation of this integral is consistent with the above theory and results from computations in [Gelbart] (in particular the relation (**) below) by a fairly direct argument. This argument, however, is not entirely obvious and has not been explained before. Therefore we shall sketch it below. We should also note that the complex analogue of the relation (**) below first appeared in [Stein].

Let u and v denote arbitrary vectors in the space of $\pi$. Then the integral

$$\int_G f(g) < \pi(g)\,u, v > |\det g|^s\,dg$$

actually converges *for* $\mathrm{Re}(s) > \frac{1}{2}$. To see this it suffices to integrate over $G^+ = GL^+(2,R) = R_+^\times \times SL(2,R)$. But by Caucy-Schwartz and the square-integrability of $<\pi(g)\,u, v>$,

$$\int_{G^+} f(g) < \pi(g)\,u, v > |\det g|^s\,dg \leq M\left\{\int_{SL(2,R)} |f_s(x)|^2 dx\right\}^{1/2}$$

where $f_s(x) = \int_0^\infty f(xr)\,r^s \frac{dr}{r}$, so to prove our claim it suffices to establish the following:

LEMMA. *Let* $S = SL(2,R)$. *Then* $f_s \in L^2(S)$ *when* $\mathrm{Re}(s) > \frac{1}{2}$.

*Proof* (Sketch). Because

$$\int_{G^+} f(g)\,|\det g|^s\,dg = \int_S f_s(x)\,dx\ ,$$

it follows that $\|f_s\|_1 < \infty$ when $\mathrm{Re}(s) = 1+\varepsilon, \varepsilon > 0$. On the other hand, $\|f_s\|_\infty < \infty$ whenever $\mathrm{Re}(s) = \varepsilon > 0$. So combining these facts with the Hadamard 3-line theorem we conclude that for $\mathrm{Re}(s) > \frac{1}{2}$,

$$\int_S f_s(x)\phi(x)\,dx \leq A\|\phi\|_2$$

for all $\phi$ in a dense subspace of $L^2(S)$. The desired result then follows from the converse to Holder's inequality. □

Now set

$$\zeta(\hat{f}, \check{\pi}, 2-s) = \int \hat{f}(g) < \check{\pi}(g)u, v > |\det g|^{2-s}\,dg$$

where $\check{\pi}(g)$ denotes the discrete series representation $\pi(g^{-1})$. By the above argument this integral converges for $\mathrm{Re}(s) < \frac{3}{2}$.

THEOREM. *The zeta-function* $\zeta(f, \pi, s)$, *initially defined for* $\mathrm{Re}(s) > \frac{1}{2}$, *continues meromorphically to the whole* s*-plane and satisfies the functional equation*

$$\zeta(\hat{f}, \check{\pi}, 2-s) = m_\pi(s)\zeta(f, \pi, s)$$

*where*

$$m_\pi(s) = i^{n+1}(2\pi)^{s+it-1}\frac{\Gamma\left(\frac{n+(2-s)-it}{2}\right)}{\Gamma\left(\frac{n+s+it}{2}\right)}$$

*if* $\pi$ *is indexed by the parameters* n *and* it. *(Cf. Remark 4.6.)*

*Proof.* The integrals defining $\zeta(f, \pi, s)$ and $\zeta(\hat{f}, \check{\pi}, 2-s)$ converge simultaneously in the strip

$$\frac{1}{2} < \mathrm{Re}(s) < \frac{3}{2}$$

and so define analytic functions there. By Theorem 3 of [Gelbart]

$$(**) \qquad \int_G \hat{f}(x)\check{\pi}(g)\,|g|\,dg = m_\pi(1) \int_G f(g)\pi(g)\,|g|\,dg$$

which says that

$$\zeta(\hat{f}, \check{\pi}, 2-s) = m_\pi(s)\zeta(f, \pi, s)$$

when $\mathrm{Re}(s) = 1$. By analytic continuation then this relation also holds throughout the strip $\frac{1}{2} < \mathrm{Re}(s) < \frac{3}{2}$ and hence everywhere since $m_\pi(s)$ is obviously meromorphic in $\mathbf{C}$. □

An alternate approach to this theorem due to Godement is described in Section 9 or [Godement-Jacquet]. See also Section 13 of [Jacquet-Langlands]. The special case of the trivial representation is discussed in [Weil 6] as well as in the above references.

FURTHER REFERENCES

*A.* A good reference for Hecke's theory of L-series with grossen-character is [Godement 3]. Tate's thesis appears in [Cassels-Fröhlichs] but is also discussed in [Lang] and [Goldstein].

*B.* An essential feature of Jacquet-Langlands' theory is that it puts on a completely equal footing the real analytic wave-forms of Maass and the classical holomorphic cusp forms of Hecke's theory. This feature as well as the generality of an arbitrary base field is shared by the elementary theory of [Weil 2] which runs parallel to Jacquet-Langlands theory *without* using the theory of group representations.

C. The results quoted here are (sometimes implicitly) proved in [Jacquet-Langlands].

The fundamental local theorem on the existence and uniqueness of Whittaker models (Theorem 6.8) was used by Jacquet-Langlands to classify all irreducible admissible representations as follows.

For each $W \in W(\pi_v)$ set

$$f_W(x) = W\left(\begin{bmatrix} x & 0 \\ 0 & 1 \end{bmatrix}\right).$$

The resulting space of functions on $F_v^\times$ is called the *Kirillov space* of $\pi_v$ and its properties are described by the following reformulation of the existence and uniqueness theorem for local Whittaker models.

THEOREM. *Every irreducible admissible representation* $\pi_v$ *of* $G_v$ *(of infinite dimension) is equivalent to a unique representation acting in a space of locally constant functions on* $F_v^\times$ *compactly supported in* $F_v$ *and such that*

(6.54) $$\pi_v\left(\begin{bmatrix} a & b \\ 0 & 1 \end{bmatrix}\right)f(x) = \tau_v(bx)f(ax) \quad (a \in F_v^\times,\ b \in F_v) .$$

*This space* $K(\pi_v)$ *furthermore contains the Schwartz-Bruhat space* $S(F_v^\times)$ *as a subspace of codimension at most two.*

One passes from the Kirillov model to the Whittaker model by setting

$$W_f(g) = (\pi_v(g)f)(1) .$$

The classification of the irreducible admissible (infinite dimensional) representation of $G_v$ then takes the form that $\pi_v$ is *supercuspidal* if $K(\pi_v) = S(F_v^\times)$, is *special* if $S(F_v^\times)$ has codimension 1, and belongs to the *principal* series otherwise (i.e. if $S(F_v^\times)$ has codimension two in $K(\pi_v)$). *Note that the degree of* $L(s,\pi_v)$ *is at most the codimension of* $S(F_v^\times)$ *in* $K(\pi_v)$.

We remark also that by virtue of (6.54) an irreducible representation $\pi_v$ is completely determined by knowledge of the single operator $\pi\left(\begin{bmatrix} 0 & 1 \\ -1 & 0 \end{bmatrix}\right)$ once one *knows* the central character of $\pi_v$ and its space $K(\pi_v)$.

The terminology for $K(\pi_v)$ seems apt because it was Kirillov who first proved that (over *any* local field) every irreducible unitary representation $\pi_v$ of $G_v$ remains irreducible upon restriction to the subgroup

$$\left\{\begin{bmatrix} a & b \\ 0 & 1 \end{bmatrix}\right\}$$

and on this subgroup is of the form (6.54). In the real case such facts had already been exploited in [Kunze-Stein].

A very readable account of Theorem 6.8 is to be found in [Godement]. Using the theory of spherical functions a completely different argument for uniqueness has recently been described by R. Howe.

## §7. THE CONSTRUCTION OF A SPECIAL CLASS OF AUTOMORPHIC FORMS

The problem of constructing automorphic forms is a difficult one. Of course one knows that *holomorphic* cusp forms of integral weight exist and one knows how to construct them using Poincaré series (cf. Example B(ii)). On the other hand one knows essentially nothing about which *real* analytic cusp forms exist let alone how to construct them. In this direction the pioneering work is due to Maass. Maass reduced the problem of constructing real analytic wave forms to the study of certain L-series with functional equation attached to real (as opposed to complex) quadratic fields. In his work (as well as in related works by Hecke) theta series attached to grossencharacters played a fundamental role.

The purpose of this section is to explain how Jacquet-Langland's theory considerably generalizes and smooths out the techniques of (Hecke and) Maass. The result is the construction of a wide (but still special) class of cusp-forms indexed by certain grossencharacters of global fields. The representations in question are constructed using the *representation*-theoretic formulation of theta-functions due to Weil. The fact that they actually define cusp forms will follow from a study of their L-functions.

In the first subsection below we summarize some salient features of the so-called Weil representation and its applicability to representation theory. We especially have in mind the problem of constructing supercuspidal representations. Because the Weil representation is also used in Section 10 in a somewhat different context we develop it in greater generality than is required at present.

In Subsection B we apply the Weil representation to the construction of cusp forms as alluded to above. Then in Subsection C this construction is specialized to provide explicit examples of non-zero real analytic cusp forms of specified level N and eigenvalue s. Subsection D sketches the connections between this construction (and its generalizations) and certain natural conjectures in non-abelian class field theory.

## A. The Weil Representation

Let F denote a local field and G the group GL(2, F). In this context the Weil representation is a representation of G which one canonically associates to any quadratic form defined over F.

For our purposes it will suffice to consider forms q of the following type:

(i) q is the norm form of a separable quadratic extension L of F; or

(ii) q is the reduced norm of the unique quaternion division algebra D over F.

In either case let $x \to x^\sigma$ denote the canonical involution of $V = L$ or D over F. Then

$$q(x) = xx^\sigma$$

for all x in V and

$$x + x^\sigma = tr(x)$$

where (tr) denotes the trace (or reduced trace) from $V = L$ (or D) to F.

Now fix a non-trivial additive character $\tau$ of F. Since $(x,y) \to tr(xy)$ is a non-degenerate bilinear form on V, V can be identified with its dual by the pairing

$$\langle x, y \rangle = \tau(tr(xy)) .$$

The Fourier transform $\hat{\Phi}$ of each $\Phi$ in the Schwartz-Bruhat space S(V) is by definition

$$\hat{\Phi}(x) = \int_D \Phi(y) \langle x, y \rangle dy \tag{7.1}$$

where Haar measure dy is normalized so that $(\hat{\Phi})^\wedge(x) = \Phi(-x)$.

To describe the Weil representation corresponding to the pair (q,V) it is necessary to introduce an important invariant for V. First put

(7.2) $$f(x) = \tau(q(x)) .$$

By [Weil 3] there exists a constant $\gamma$ depending only on D (and $\tau$) such that

(7.3) $$(\Phi^* f)\hat{}\,(x) = \gamma f^{-1}(x^\sigma)\hat{\Phi}(x)$$

for all $\Phi \in S(V)$. In particular, $\gamma = -1$ if $V = D$. If $V = L$ suffice it to say that $|\gamma| = 1$.

Now consider the representation r(s) of SL(2, F) in S(V) such that

(7.4) $$r\left(\begin{bmatrix} 1 & u \\ 0 & 1 \end{bmatrix}\right)\Phi(x) = \tau(uq(x))\Phi(x) ,$$

(7.5) $$r\left(\begin{bmatrix} a & 0 \\ 0 & a^{-1} \end{bmatrix}\right)\Phi(x) = \omega(a)\,|a|_V^{\frac{1}{2}}\,\Phi(ax) ,$$

and

(7.6) $$r\left(\begin{bmatrix} 0 & 1 \\ -1 & 0 \end{bmatrix}\right)\Phi(x) = \gamma\hat{\Phi}(x^\sigma) .$$

Here $\omega$ denotes the non-trivial character of $F^\times/\mathrm{Norm}(V^\times)$ if $V = L$ and the trivial character of $F^\times$ if $V = D$. Since elements of the form

(7.7) $$\begin{bmatrix} a & 0 \\ 0 & a^{-1} \end{bmatrix}, \begin{bmatrix} 1 & u \\ 0 & 1 \end{bmatrix}, \text{ and } \begin{bmatrix} 0 & 1 \\ -1 & 0 \end{bmatrix}$$

generate SL(2, F) there will be *at most* one representation of SL(2, F) satisfying (7.4)-(7.6). That there is one such representation (call it r(s)) is a result of Weil, Shalika, and Tanaka. To prove it one has simply to verify that the mapping

$$s \to r(s)$$

specified by (7.4)-(7.6) preserves all relations between the generators (7.7).

The representation r(s) just described is *"the" Weil representation of* SL(2, F) associated to the pair (V, q) and it may depend on the choice of character $\tau$. However, if $a \in F^\times$, and $\tau_a(x) = \tau(ax)$, then the representation $r_a(s)$ corresponding to $\tau_a$ is related to r(s) through the formula

(7.8) $$r_a(s) = \left(\begin{bmatrix} a & 0 \\ 0 & 1 \end{bmatrix} s \begin{bmatrix} a^{-1} & 0 \\ 0 & 1 \end{bmatrix}\right).$$

If $b \in V^x$ define operators $\lambda(b)$ and $\rho(b)$ in $S(V)$ by

(7.9) $$\lambda(b)\Phi(x) = \Phi(b^{-1}x)$$

and

(7.10) $$\rho(b)\Phi(x) = \Phi(xb) .$$

Then if $a = q(b)$, a straightforward computation shows that

(7.11) $$r_a(s)\lambda(b^{-1}) = \lambda(b^{-1})r(s)$$

and

(7.12) $$r_a(s)\rho(b) = \rho(b)r(s) .$$

In particular $r(s)$ *and* $r_a(s)$ *are equivalent* if $a$ belongs to the subgroup $q(V^x)$ of $F^x$. (So since $q(V^x) = F^x$ if $V = D$ and $F$ is not $\mathbf{R}$, the Weil representation in this case is independent of the choice of $\tau$.) Furthermore from (7.11) and (7.12) it is clear that both $\lambda(b)$ *and* $\rho(b)$ *commute with* $r(s)$ *if* $q(b) = 1$.

REMARK 7.1. If $S(V)$ is given its usual topology, $r(s)$ is continuous. Furthermore $r(s)$ extends to a unitary representation of $SL(2,F)$ on $L^2(V)$.

REMARK 7.2 (*Concerning the General Construction of Weil*). The representations $r(s)$ corresponding to the forms $q$ on $L$ and $D$ represent special cases of a very general construction devised by Weil, Segal and Mackey. In Weil's theory a "big" representation is canonically attached to the "abstract symplectic group" of any locally compact abelian group $G$. The connection between this set-up and the above representations is that for certain special choices of $G$ the group $SL(2,F)$ imbeds naturally in the symplectic group of $G$ and $r(s)$ appears as the restriction of this big representation to $SL(2,F)$.

In rough detail, the situation is as follows. Fix a duality $\langle \cdot,\cdot \rangle$ between $G$ and its dual $G^*$. For each $w = (u,u^*)$ in $G \times G^*$, let $U(w)$ be the unitary operator in $L^2(G)$ taking $\phi(x)$ to $\phi'(x) = \langle x, u^* \rangle \phi(x+u)$. Then $U(w)$ is a projective representation of $G \times G^*$ with multiplier $F(w_1, w_2) = \langle u_1, u_2^* \rangle$ if $w_i = (u_i, u_i^*)$, i.e. $U(w_1)U(w_2) = U(w_1+w_2)F(w_1,w_2)$.

Let $\bar{A}(G)$ denote the group of all unitary operators in $L^2(G)$ of the form $tU(w)$ with $|t| = 1$ and $\bar{B}_0(G)$ its normalizer in $U(L^2(G))$. Since $\bar{A}(G)$ is isomorphic to the group $A(G)$ of all pairs $w, t$ with $w \in G \times G^*$, $|t| = 1$, and $(w_1, t_1)(w_2, t_2) = (w_1+w_2, F(w_1, w_2)t_1t_2)$, each member of $\bar{B}_0(G)$ induces an automorphism of $A(G)$ which has the form $(\omega, t) \to (\omega\sigma, tg(w))$ where $\sigma$ is an automorphism of $G \times G^*$ and $g$ is a continuous function from $G \times G^*$ to the circle group related to $\sigma$ by the identity

$$F(w_1\sigma, w_2\sigma) = F(w_1, w_2)g(w_1+w_2)/g(w_1)g(w_2) .$$

Conversely any pair $\sigma, g$ with $\sigma$ and $g$ so related defines an automorphism of $A(G)$ and the group of all such is denoted by $B_0(G)$.

Now let $p_0$ denote the homomorphism of $\bar{B}_0(G)$ into $B_0(G)$ just described. Weil proves that $p_0$ is surjective with kernel equal to the group of constant multiples of the identity. *Consequently one obtains a projective representation of* $B_0(G)$ *in* $L^2(G)$. The symplectic group $Sp(G)$ associated to $G$ is the group of automorphisms of $G \times G^*$ preserving the form $F(w_1, w_2)F(w_2, w_1)^{-1}$. Equivalently $\sigma = \begin{bmatrix} \alpha & \beta \\ \gamma & \delta \end{bmatrix}$ is symplectic if $\sigma\sigma^I = I$ where $\sigma^I = \begin{bmatrix} \delta^* & -\beta^* \\ -\gamma^* & \alpha^* \end{bmatrix}$. (Here $\alpha, \beta, \gamma$ and $\delta$ are homomorphism of $G$ into $G$, of $G$ into $G^*$, of $G^*$ into $G$, and $G$ into $G^*$ respectively, $w\sigma = (u, u^*)\sigma = (u, u^*)\begin{bmatrix} \alpha & \beta \\ \gamma & \delta \end{bmatrix}$, and $\alpha^*$ denotes the dual of $\alpha$.) Since $Sp(G)$ injects in $B_0(G)$ through the homomorphism $\begin{bmatrix} \alpha & \beta \\ \gamma & \delta \end{bmatrix} \to (\sigma, g_\sigma)$ where

$$g_\sigma(u, u^*) = \langle u, 2^{-1}u\alpha\beta^* \rangle \, \langle 2^{-1}u^*\gamma\delta^*, u^* \rangle \, \langle u^*\gamma, u\beta \rangle$$

*there is associated to* $Sp(G)$ *a natural (possibly projective) representation in* $L^2(G)$. *This is the Weil representation attached to* $G$.

To obtain the representation $r(s)$ of Remark 7.1 take $G$ to be the additive group of $V$ and fix the self-duality

$$\langle u, v \rangle = \tau(2B_q(u, v))$$

if $B_q$ denotes the bilinear form associated to $(q,V)$. Since each $a \in F$ defines a homomorphism of $G = V^+$ (by scalar multiplication), and since $\begin{bmatrix} \alpha & \beta \\ \gamma & \delta \end{bmatrix}$ in SL(2, F) satisfies $\sigma\sigma^I = I$, each element of SL(2, F) imbeds homomorphically in $B_0(G)$. By restriction there results an (ordinary) representation of SL(2, F) which coincides with the special representation r(s) described earlier.

Before returning (finally) to our discussion for SL(2, F) let us note that the assertion "$p_0$ is surjective" is essentially an analogue (for local fields) of the Stone-von Neumann Uniqueness Theorem for representations of the Heisenberg commutation relations. Indeed this theorem asserts that (up to equivalence) there is exactly *one* irreducible representation of $G \times G^*$ with multiplier F.

To describe *the Weil representation for* GL(2, F) corresponding to the pair (q,V) one introduces the group $G_+$ consisting of n in GL(2, F) with determinant in $q(V^x)$. This group has index 2 or 1 in GL(2, F) according if $V = L$ or $V = D$ (with $F \neq R$).

LEMMA 7.3. *The representation* r(s) *of* SL(2, F) *on* S(V) *extends to a representation* r(n) *of* $G_+$ *such that*

$$r\left(\begin{bmatrix} a & 0 \\ 0 & 1 \end{bmatrix}\right)\Phi(x) = \Phi(xh) \tag{7.13}$$

*if* $a = q(h)$ *belongs to* $q(V^x)$.

*Proof.* Let C denote the group of matrices

$$\left\{\begin{bmatrix} a & 0 \\ 0 & 1 \end{bmatrix}\right\}$$

with a in $q(V^x)$ and observe that $G_+$ is the semi-direct product of G and SL(2, F). Clearly (7.13) defines a continuous representation of H on $S(V^x)$. Therefore to prove the lemma it suffices to verify the relation

$$(7.14)\qquad r\left(\begin{bmatrix} a & 0 \\ 0 & 1 \end{bmatrix} s \begin{bmatrix} a^{-1} & 0 \\ 0 & 1 \end{bmatrix}\right) = r\left(\begin{bmatrix} a & 0 \\ 0 & 1 \end{bmatrix}\right) r(s)\, r\left(\begin{bmatrix} a^{-1} & 0 \\ 0 & 1 \end{bmatrix}\right).$$

But the left side of (7.14) is simply $r_a(s)$ and

$$r\left(\begin{bmatrix} a & 0 \\ 0 & 1 \end{bmatrix}\right)\Phi(x) = \rho(h)\Phi(x) .$$

Consequently (7.14) and the lemma follow from (7.12). □

To emphasize the possible dependence of $r(n)$ on $\tau$ we denote it by $r_\tau(n)$. Now let $r(g)$ denote the representation of $GL(2,F)$ induced from $r_\tau(n)$. This $r(g)$ is "*the Weil representation of* $GL(2,F)$ (associated to the pair $(q,V)$) and *it is independent of the choice* of character $\tau$. (Indeed $r(g)$ is obtained by letting $GL(2,F)$ act by right translation on the space of functions $\phi$ on $G$ (with values in the space of $r_\tau(n)$) which satisfy

$$\phi(ng) = r_\tau(n)\phi(g)$$

for all $n \in G_+$. If $a \in F^x$ and we define $\phi^a(g) = \phi\left(\begin{bmatrix} a & 0 \\ 0 & 1 \end{bmatrix} g\right)$ then $\phi^a$ is 0 iff $\phi$ is, and

$$\begin{aligned}\phi^a(ng) &= \phi\left(\begin{bmatrix} a & 0 \\ 0 & 1 \end{bmatrix} ng\right) \\ &= r_\tau\left(\begin{bmatrix} a & 0 \\ 0 & 1 \end{bmatrix} n \begin{bmatrix} a^{-1} & 0 \\ 0 & 1 \end{bmatrix}\right)\phi^a(g) .\end{aligned}$$

That is, replacing the space $\{\phi\}$ by the space $\{\phi^a\}$, we obtain an equivalent representation of $G$ induced from the representation

$$n \to (r_\tau)_a(n)$$

of $G_+$. So since $a$ is arbitrary it follows that $r(g)$ is independent of $\tau$.)

We close this subsection by describing how the representations already introduced allow one to construct a wide class of irreducible admissible representations of $GL(2,F)$. This connection between Weil's construction and irreducible representation theory was first made explicit in [Shalika] and [Tanaka] and then completely exploited in [Jacquet-Langlands]. The crucial point is a relation of the type (7.11) or (7.12). Such a relation asserts that the regular representation of $V^x$ in S(V) commutes with r(s). This in turn implies that the Weil representation is highly reducible and in fact decomposes into irreducibles according to the representation theory of $V^x$.

So suppose $\pi'$ is an irreducible (finite-dimensional) representation of $V^x = G'$ on a vector space H defined over C. Taking the tensor product of r(s) with the trivial representation of SL(2, F) on H one obtains a representation on

$$S(V) \otimes_C H$$

which we again denote by r. An element in $S(V) \otimes H$ will be regarded as a function on V taking values in H whose coordinate entries (with respect to some basis of H) are Schwartz-Bruhat functions on V. Because of (7.12) the subspace of functions in $S(V) \otimes H$ satisfying

$$\Phi(xh) = \pi'(h^{-1})\Phi(x) \tag{7.15}$$

for all h in $G'$ with $q(h) = 1$ is immediately seen to be invariant for r(s). The resulting subrepresentation of r(s) will be denoted by $r_{\pi'}(s)$.

By the reasoning used to prove Lemma 7.3 it can be shown that $r_{\pi'}(s)$ extends to a representation of $G_+$ satisfying

$$r_{\pi'}\left(\begin{bmatrix} a & 0 \\ 0 & 1 \end{bmatrix}\right)\Phi(x) = |h|_V^{1/2}\,\pi'(h)\Phi(xh) \tag{7.16}$$

if $a = q(h)$ belongs to $q(V^x)$. Furthermore

$$r_{\pi'}\left(\begin{bmatrix} a & 0 \\ 0 & a \end{bmatrix}\right) = \omega(a)\chi_{\pi'}(a)I$$

if a is arbitrary in $F^x$ and $\chi_{\pi'}$ is the central character of $\pi'$. If $\pi'$ is unitary (and H is a Hilbert space) then $r_{\pi'}$ *can be extended to a unitary representation of* $G_+$ in $L^2(V, \pi')$, the closure of $S(V, \pi')$ in the Hilbert space of square integrable functions from V to H with norm $\|\Phi\|^2 = \int_V \|\Phi(x)\|^2 dx$.

Now consider the representation of $GL(2,F)$ induced from $r_{\pi'}(n)$ and denote it again by $r_{\pi}$. Of course $r_{\pi'}(g)$ is independent of the choice of character $\tau$ since it is a component of the Weil representation $r(g)$ of $GL(2,F)$. What is remarkable is that $r_{\pi'}(g)$ is always admissible and (essentially) always an irreducible representation. The equivalence class of $r_{\pi'}(g)$ depends on the choice of $\pi'$ and on whether $V$ equals $L$ or $D$.

THEOREM 7.4 ($V$ is a separable quadratic extension of $F$).

(i) *If there is no character* $\delta$ *of* $F^{\times}$ *such that* $\pi' = \delta \circ q$ *then* $r_{\pi'}(g)$ *is a super cuspidal representation of* $GL(2,F)$.

*In fact by varying such* $\pi'$ *as* $V$ *varies through all equivalence classes of quadratic extensions, all super cuspidal representations of* $GL(2,F)$ *may be so obtained;*

(ii) *If* $\pi' = \delta \circ q$ *for some character* $\delta$ *of* $F^{\times}$ *then* $r_{\pi'}(g)$ *is equivalent to the principal series representation* $\pi(\delta, \delta\omega)$.

REMARK 7.5. The kernel of $q$ is the norm one group $L^1$ of $L$. Thus $\pi' = \delta \circ q$ for some character $\delta$ of $F^{\times}$ if and only if $\pi'$ is trivial on $L^1$. In particular, $r_{\pi'}$ is *super cuspidal only if* $\pi'$ *is a ramified character of* $L^{\times}$. Note too that if $\pi'$ is trivial on $L^1$ it factors through $F^{\times}$ by exactly two characters, namely $\delta$ and $\delta\omega$.

THEOREM 7.6 ($V$ is the unique division quaternion algebra $D$ defined over $F$).

(i) *The representation* $r_{\pi'}(g)$ *decomposes as the direct sum of* $d = \dim(\pi')$ *mutually equivalent irreducible representations* $\pi(\pi')$ *of* $GL(2,F)$;

(ii) *Each* $\pi(\pi')$ *is absolutely cuspidal if* $d > 1$ *and special if* $d = 1$;

(iii) *All the super cuspidal and special representations of* $GL(2,F)$ *are obtained in this manner. More precisely, the map*

$$\pi' \to \pi(\pi')$$

*gives a one-to-one correspondence between the equivalence classes of (finite-dimensional) irreducible representations of* $G'$ *and the equivalence classes of special and super cuspidal representations of* $GL(2,F)$.

REMARK 7.7. In stating the theorems above we have implicitly assumed $F$ to be non-archimedean. In case $F = \mathbf{R}$, let $D$ denote the quaternion algebra over $\mathbf{R}$ identified with the set of matrices of the form $\begin{bmatrix} a & b \\ -\bar{b} & \bar{a} \end{bmatrix}$ with $a, b$ in $\mathbf{C}$. Then $q(h) = \det(h)$, identifying

$$h = a_0 + a_1\bar{i} + a_2\bar{j} + a_3\bar{k}$$

with

$$h = \begin{bmatrix} a_0 + a_1\sqrt{-1} & a_2 + a_3\sqrt{-1} \\ -(a_2 - a_3\sqrt{-1}) & a_0 - a_1\sqrt{-1} \end{bmatrix} \tag{7.17}$$

Every irreducible representation of $G' = D^{\times}$ is known to be of the form

$$\pi'(h) = q(h)^r \rho_n(h) \tag{7.18}$$

where $r \in \mathbf{C}$, and $\rho_n(h)$ is the n-th symmetric tensor power of the standard two dimensional representation $\rho_1$ of $G'$ associated with the identification (7.17). Then

$$\pi(\pi') = \sigma(\mu_1, \mu_2) \ , \tag{7.19}$$

where $\mu_1(a) = |a|^{r+n+\frac{1}{2}}$, and $\mu_2(a) = |a|^{r-\frac{1}{2}}[\operatorname{sgn}(a)]^n$, and $\sigma(\mu_1, \mu_2)$ denotes the appropriate discrete series representation of $GL(2,\mathbf{R})$ in the notation of Section 4. Furthermore every such representation of $GL(2,\mathbf{R})$ is obviously obtained.

In case $F = \mathbf{R}$ and $V = L = \mathbf{C}$ the statement of Theorem 7.4 is simplest of all. If $\pi'(z)$ is a character of $\mathbf{C}^{\times}$ *not of the form* $\chi(z\bar{z})$ with $\chi$ a quasi-character of $\mathbf{R}^{\times}$, then $\pi(\pi')$ is of the form

$$\pi'(z) = (z\,\bar{z})^r z^m \bar{z}^n \tag{7.20}$$

with $r \in \mathbf{C}$, and $m$ and $n$ two integers, one zero and the other *positive*. In this case, $r_{\pi'}$ is equivalent to the *discrete series representation* $\sigma(\mu_1, \mu_2)$, with $\mu_1\mu_2(t) = |t|^{2r} t^{m+n} \operatorname{sgn}(t)$ and $\mu_1\mu_2^{-1}(t) = t^{m+n} \operatorname{sgn}(t)$. If, on the other hand, $\pi'$ is of the form (7.20) with $n+m = 0$, i.e. if $\pi'(z)$ is of the form $\chi(z\,\bar{z})$, then $r_{\pi'}$ is equivalent to the principal series representation $\pi(\mu_1, \mu_1 \operatorname{sgn}(t))$.

SUMMING UP. Let $V$ denote a division algebra over the local field $F$ and $q$ a quadratic form on $V$. To the pair $(q,V)$ there is associated a natural representation $r$ of $GL(2,F)$ called the *Weil representation* of $G$. In case $V$ is a separable quadratic extension of $F$ and $q$ the norm form of this extension, the decomposition of $r(g)$ according to the irreducible representations $\pi'$ of $V^x$ contains supercuspidal and principal series representations. In case $V$ is the unique division quaternion algebra over $F$ and $q$ *its* norm form, the decomposition of $r(g)$ according to the irreducible representations $\pi'$ of $V^x$ contains *all* the supercuspidal and special representations of $GL(2,F)$. (In case $F = \mathbf{R}$ replace "supercuspidal" and "special" by "discrete series".) In either case there results a correspondence

$$\pi' \to \pi(\pi') \tag{7.21}$$

between irreducible representations of $G' = V^x$ and $G = GL(2,F)$.

In this section we shall henceforth restrict our attention to the correspondence (7.21) for $G' = L^x$. The case of quaternion algebras will play a fundamental role in Section 10. In this case the Weil representation corresponding to $(D,q)$ will play the role of (constructing) theta series associated to quaternary (as opposed to binary) quadratic forms.

## B. The Construction of Certain Special Representations of GL(2, A)

Let $F$ denote a global field and $L$ a separable quadratic extension of $F$. By the main theorem of Jacquet-Langlands theory the problem of constructing (generalized) cusp forms on GL(2) is reduced to the problem of constructing representations of $GL(2,A)$ whose associated L-functions satisfy certain prescribed analytic conditions. The purpose of this subsection is to explain how the Weil representation allows one to construct a wide class of such representations indexed by grossencharacters of L.

If $\lambda$ is any grossencharacter of L write

$$\lambda = \prod_w \lambda_w \tag{7.22}$$

where each $\lambda_w$ is a character of $(L_w)^x$ and w is a place of L. We shall attach to $\lambda$ a representation

$$\pi(\lambda) = \bigotimes_v \pi_v$$

of $GL(2,A(F))$ by defining $\pi_v$ for each place v of F in terms of the character (or characters) $\lambda_w$ lying over v. This must of course be done in such a way that $\pi_v$ is class 1 for almost every v.

Suppose first that v is a prime in F which splits in L with divisors w and w′. Then $L_w$ and $L_{w'}$ are isomorphic to $F_v$ and $\lambda_w$ and $\lambda_{w'}$ may be viewed as characters of $F_v^x$. In this case we set $\pi_v$ equal to the principal series (actually continuous series) representation $\pi(\lambda_w,\lambda_{w'})$.

Suppose, on the other hand, that v does not split in L but lies under the single prime w. Then $L_w$ is a genuine quadratic extension of $F_v$ and $\lambda_w$ is a character of $L_w^x$. In this case the Weil construction

$$\pi' \to \pi(\pi')$$

is applicable with $\lambda_w$ playing the role of $\pi'$. Therefore we set $\pi_v$ equal to $\pi(\lambda_w)$.

PROPOSITION 7.8. *Let the family* $\{\pi_v\}$ *be as above. Then the formula*

$$\pi(\lambda) = \bigotimes_v \pi_v \tag{7.23}$$

*defines an irreducible unitary representation of* $GL(2,A(F))$.

*Proof.* It suffices to show that $\pi_v$ is class 1 for almost every v. But by Theorem 4.23, $\pi_v$ is *not* class 1 only if one of the following possibilities obtains:

(a) $\pi_v = \pi(\lambda_w)$ is absolutely cuspidal;

(b) $\pi_v = \pi(\lambda_w, \lambda_{w'})$ with $\lambda_w$ or $\lambda_{w'}$ ramified; or

(c) $\pi_v = \pi(\delta_v, \delta_v \omega_v)$ with $\delta_v$ or $\omega_v$ ramified.

Now by Remark 7.5, (a) obtains only if $\lambda_w$ is ramified, and (c) obtains only if v ramifies in L or if $\lambda_w = \delta_v \circ q_v$ is ramified. Since these latter possibilities occur for only *finitely* many places the proof is complete. □

To show that $\pi(\lambda)$ is a cusp form we must analyze its corresponding L-function. So let $L(s, \pi(\lambda))$ denote the L-function attached to $\pi(\lambda)$ by Jacquet-Langlands' theory. To prove that $L(s, \pi(\lambda))$ satisfies the conditions of Theorem 6.18 (and hence that $\pi(\lambda)$ occurs in some $R_0^\psi$) we must show that

$$L(s, \chi \otimes \pi(\lambda)) = \prod_v L(s, \chi_v \otimes \pi_v) \tag{7.24}$$

is entire, bounded in vertical strips, and such that the functional equation (6.44) obtains for every grossencharacter $\chi$ of F.

To this end we recall the *idele norm* map from the idele class group of L to the idele class group of F. This map is defined by

$$N(\beta) = N((\beta_w)) = (a_v) = a \ . \tag{7.25}$$

if $a_v = \prod_{w|v} N(\beta_w)$. Through it *we can extend the grossencharacter* $\chi$ *of* F *to a grossencharacter* $\Pi \chi_w$ *of* L.

Now let $L(s, \lambda\chi)$ denote the Hecke L-series attached to the "automorphic form" $\lambda\chi$ of $L^\times = GL(1)$. The analytic properties of this L-function are well known. Therefore *our plan of attack will be to compare* $L(s, \chi \otimes \pi(\lambda))$ *with* $L(s, \lambda\chi)$. Recall that

$$L(s, \lambda\chi) = \prod_w L(s, \lambda_w \chi_w) \tag{7.26}$$

where

$$L(s,\lambda_w\chi_w) = (1-\lambda_w\chi_w(\tilde{\omega}_w)N(\tilde{\omega}_w)^{-s})^{-1} \tag{7.27}$$

if $\lambda_w\chi_w$ is unramified, and equals one otherwise. Actually in (7.26) the product is taken over the infinite places of L as well. If $L_w = \mathbf{R}$,

$$L(s,\lambda_w\chi_w) = \pi^{-\frac{1}{2}(s+it_w+m_w)}\Gamma\left(\frac{s+it_w+m_w}{2}\right) \tag{7.28}$$

for $\lambda_w\chi_w(x) = [\operatorname{sgn}(x)]^{m_w}|x|^{it_w}$, $m_w = 0$ or $1$. If $L_w = \mathbf{C}$,

$$L(s,\lambda_w\chi_w) = (2\pi)^{-(s+t_w+m_w+n_w)}\Gamma(s+t_w+m_w+n_w) \tag{7.29}$$

for $\lambda_w\chi_w(z) = |z|^{it_w}z^m\bar{z}^n$, $\inf(n,m) = 0$.

LEMMA 7.9.

$$L(s,\chi\otimes\pi(\lambda)) = L(s,\lambda\chi) \tag{7.30}$$

*Proof.* As both sides of (7.30) are defined by infinite products it will suffice to verify the equality "place by place." *Caution*: The left side is a product over the primes *of* L (with Euler factors of degree at most 1) and the right side is a product over the primes *of* F (with Euler factors of degree at most 2).

Suppose first that v splits in L and lies under w and w′. Then $\pi_v = \pi(\lambda_w,\lambda_{w'})$, and so by Theorem 6.15,

$$L(s,\chi_v\otimes\pi_v) = L(s,\chi_w\lambda_w)L(s,\chi_{w'}\lambda_{w'}) \,. \tag{7.31}$$

if v is finite. If v is infinite the same result holds by Theorem 6.16.

Next suppose that v is inert and $w|v$. Then $L_w$ is an *unramified* quadratic extension of $F_v$. If $\chi_w\lambda_w$ is ramified then $L(s,\chi_v\otimes\pi_v)$ is 1 whether $\pi_v$ is supercuspidal or not. Therefore the desired equality

$$L(s,\chi_v\otimes\pi_v) = L(s,\chi_w\lambda_w) \tag{7.32}$$

obtains by default. If, on the other hand, $\chi_w\lambda_w$ is unramified, then $\pi_v$ cannot be supercuspidal by Remark 7.5. Since $N(\tilde{\omega}_w) = \tilde{\omega}_v^2$ (where $\tilde{\omega}_w$ denotes a local uniformizing variable at $w$)

$$\begin{aligned}
L(s,\chi_w\lambda_w) &= (1-\chi_w\lambda_w(w)N(\tilde{\omega}_w)^{-s})^{-1} \\
&= (1-\delta_v(\tilde{\omega}_v^2)\chi_v(\tilde{\omega}_v^2)N(\tilde{\omega}_v)^{-2s})^{-1} \\
&= (1-\delta_v(\tilde{\omega}_v)\chi_v(\tilde{\omega}_v)N(\tilde{\omega}_v)^{-s})^{-1}(1+\delta_v(\tilde{\omega}_v)\chi_v(\tilde{\omega}_v)N(\tilde{\omega}_v)^{-s})^{-1}
\end{aligned}$$

if $\lambda_w = \delta_v \circ q$. Since $v$ is inert, $\omega_v(\tilde{\omega}_v) = -1$.

Thus

$$\begin{aligned}
L(s,\lambda_w\lambda_w) &= (1-\delta_v(\tilde{\omega}_v)\chi_v(\tilde{\omega}_v)N(\tilde{\omega}_v)^{-s})^{-1}(1-\delta_v(\tilde{\omega}_v)\omega_v\chi_v(\tilde{\omega}_v)N(\tilde{\omega}_v)^{-s})^{-1} \\
&= L(s,\chi_v\otimes\pi(\lambda_v,\lambda_v\omega_v)) \\
&= L(s,\chi_v\otimes\pi_v)
\end{aligned}$$

and (7.32) again obtains.

The case when $v$ is infinite or ramified is left to the reader. □

LEMMA 7.10. *$L(s,\chi\otimes\pi(\lambda))$ is entire, bounded in vertical strips, and satisfies the requisite functional equation for all $\chi$ provided there is no grossencharacter $\mu$ of $F$ such that $\lambda = \mu\circ N$ with $N$ as in 7.25.*

*Proof.* By Lemma 7.9,

$$L(s,\chi\otimes\pi(\lambda)) = L(s,\chi\lambda)\ . \tag{7.33}$$

Therefore $L(s,\chi\otimes\pi(\lambda))$ satisfies the requisite functional equation since $L(s,\chi\lambda)$ does by Hecke's abelian theory. (Thus far we have made no use of the hypothesis on $\lambda$.) It remains however to check that $L(s,\chi\lambda)$ is entire and bounded in vertical strips.

According to Hecke's theorem for $L(s,\chi\lambda)$ as defined by (7.26) (cf. [Weil 4], Theorem 5, p. 133) $L(s,\chi\lambda)$ is entire *if* $\omega\lambda$ is a "non-trivial" grossencharacter of $L$, i.e. if

(7.34) $$\chi\lambda(a) \not\equiv |a|^r \quad \text{for} \quad a \in A^x(L) \ .$$

But (7.34) is equivalent to the condition

$$\lambda(a) \not\equiv \chi(a)|a|^r \ .$$

So since $\chi$ and $|\cdot|^r$ both come from characters of the idele class group of F ($\chi$ by definition and $|a|^r = \prod_w \psi_w$ because $\psi_w = \psi_{w'}$ whenever $w, w'|v$) it follows that condition (7.34) is *implied* by our hypothesis on $\lambda$. Consequently $L(s,\chi\lambda)$ (and hence $L(s,\chi\otimes\pi(\lambda))$) is entire.

That $L(s,\chi\lambda)$ is bounded in vertical strips follows from the integral representation for the global zeta-function $\zeta(f,\chi\lambda,s)$ and the fact that $\zeta(f,\chi\lambda,s)$ essentially equals $L(s,\chi\lambda)$ for an appropriate choice of f. □

Summing up, we have:

THEOREM 7.11. *Let* L *denote any separable quadratic extension of a global field* F. *The map*

$$\lambda \to \pi(\lambda)$$

*just described defines a correspondence between grossencharacters of* L *(i.e. automorphic forms on* GL(1) *over* L) *and irreducible representations of the adele group of* GL(2) *over* F. *If* $\lambda$ *is not of the form* $\mu\circ N$ *for any grossencharacter* $\mu$ *of* F *then* $\pi(\lambda)$ *is actually a cusp form on* GL(2).

Suppose $\psi$ denotes the central character of $\pi(\lambda)$. From the theorem above it follows that there is a subspace of functions in $L_0^2(G_F\backslash G_A,\psi)$ which is equivalent as a $G_A$-module to $\pi(\lambda)$. In particular, if $F = \mathbf{Q}$, these functions describe classical holomorphic or real analytic cusp forms of level equal to the conductor of $\pi(\lambda)$. The drawback of the elegant proof of Theorem 7.11 just sketched is that it does not construct these functions directly. This situation is in contrast to the classical approach involving theta series alluded to earlier. Since the constructive theme of this classical approach is desirable from several points of view we shall briefly sketch its modern formulation below.

The direct construction of cusp forms in the space of $\pi(\lambda)$ (in the generality of Theorem 7.11) is due to [Shalika-Tanaka]. Their point of departure is a *global* analogue of the Weil representation which may be described as follows. (Cf. [Weil 3], Chapter III.)

As before let L denote a separable quadratic extension of F, and $\tau = \prod_v \tau_v$ an additive character of F. If the place v of F lies under the single prime w of L, let $L_v$ denote the completion $L_w$. On the other hand, if both w and w′ in L divide v, put $L_v$ equal to the direct sum $L_w \oplus L_{w'}$. In either case, $L_v$ will be a semi-simple commutative algebra of rank two over F equipped with a quadratic form $q_v$ given by the relative norm from $L_v$ to $F_v$. (In case $L_v$ is a quadratic extension of $F_v$, $q_v$ is the usual norm form; in case $L_v = L_w \oplus L_{w'}$, $q_v = xx^\sigma$ with $x^\sigma = (b,a)$ if $x = (a,b)$.)

By the general theory of Weil there is a constant $\gamma_v$ (of absolute value 1) such that the formulas (7.4)-(7.6) define a *representation* of $SL(2,F_v)$ on $S(L_v)$ (the *local Weil representation*). This representation extends to a unitary representation $r_v$ of $L^2(L_v)$ which coincides with the Weil representation introduced in Subsection A if $L_v$ is a quadratic extension. If $L_v$ is an unramified quadratic extension *or* is isomorphic to $L_w \oplus L_{w'}$ then $r_v$ is class 1 (a $K_v$-fixed vector being the characteristic function of the maximal order of $L_v$). Thus we get a representation r of $SL(2,A(F))$ in $S_0(L_A) = \otimes S(L_v)$, a subspace of the Schwartz-Bruhat space $S(L_A)$, by setting

$$r(s)\Big(\prod_v \Phi_v\Big) = \prod_v r_v(L_v)(s_v)\Phi_v \ . \tag{7.35}$$

The representation r(s) is called *the global Weil representation* (attached to L (and $\tau$)). It can be shown to extend to $S(L_A)$ in such a way that the mapping $(s,\Phi) \to r(s)\Phi$ of $SL(2,A) \times S(L_A)$ into $S(L_A)$ is continuous. Not surprisingly it also extends to a unitary representation in $L^2(L_A)$ such that

$$r\left(\begin{bmatrix} 0 & 1 \\ -1 & 0 \end{bmatrix}\right)\Phi(x) = \gamma\hat{\Phi}(x^{\sigma}) \tag{7.36}$$

and

$$r\left(\begin{bmatrix} 1 & u \\ 0 & 1 \end{bmatrix}\right)\Phi(x) = \tau(bN(u))\Phi(x) \tag{7.37}$$

with $\gamma = \Pi\gamma_v$.

Now (at last) we can explain what this representation has to do with theta-series and the explicit construction of cusp forms. In [Weil 3] it is shown that the series

$$\Theta(\Phi) = \sum_{\xi \in L} \Phi(\xi) \tag{7.38}$$

converges uniformly on any compact subset of $S(L_A)$. Therefore the function $\Theta(r(s)\Phi)$, for each $\Phi$ in $S(L_A)$, is a continuous function on SL(2,A) which is left SL(2, F) invariant by virtue of (7.36) and (7.37). The resulting space of functions

$$\Theta(r(s)\Phi) = \sum_{\xi \in L} r(s)\Phi(\xi) \tag{7.39}$$

comprises a space of *generalized* theta functions.

To obtain cusp forms on GL(2) (actually SL(2)) one decomposes this space of theta functions according to the characters $\lambda$ of the relative norm one group $H^1$ of $L_A$ (which are trivial on $H^1_F$). This decomposition runs parallel to the decomposition of the local Weil representations $r_v(L_v)$ according to the local characters $\lambda_w$.

More precisely, consider the functions

$$\Theta_{\Phi}(\lambda)(s) = \int_{H^1\backslash H^1_A} \lambda(n)\left(\sum_{\xi \in L} r(s)\Phi(\xi n)\right) dn \tag{7.40}$$

obtained by projecting the theta-functions $\Sigma\, r(s)\Phi(\xi)$ onto the space of theta-functions which transform under $L^1_A$ according to $\lambda$. These functions are cusps forms on SL(2) and their space realizes the (essentially) irreducible unitary representation $\pi(\lambda, V) = \otimes r_v(\lambda_v, L_v)$, the components $r_v(\lambda_v, L_v)$ of $r_v$ being defined in the obvious way.

In Section 10 we shall describe a similar construction of cusp forms for GL(2) associated to the Weil representation of a global *quaternion algebra* D. In this case the theta functions that arise correspond to quaternary quadratic forms. Since the relation between these theta functions and certain classical theta series will be discussed in more detail then, we shall refrain now from making any further remarks about the theta-functions described by (7.40) except to say that they generalize the theta-series with grossencharacter considered in [Maass].

C. An Explicit Example

It has already been remarked several times that very little is known concerning which *continuous* series representations of $SL(2, \mathbf{R})$ occur *discretely* in $L^2(\Gamma_0(N)\backslash SL(2, \mathbf{R}))$. We do know that the sequence of real numbers $\{s_j\}$ parameterizing such representations indexes non-zero spaces of real analytic cusp forms. However, the arithmetic significance of this sequence is not known.

The purpose of this subsection is to apply the results of Jacquet-Langlands theory to obtain *explicit* examples of such representations for various congruence subgroups. Although we shall supply details for but one example it will be clear from the context how to construct others. Unfortunately this construction tells us little about the nature of the sequence $\{s_j\}$ for a *fixed* given level N.

Let L denote a quadratic extension of $\mathbf{Q}$. Our goal is to construct a grossencharacter $\lambda$ of L such that $\pi(\lambda)$ is a cusp form for $GL(2, \mathbf{A}(\mathbf{Q}))$ with the property that its infinite component is equivalent to some continuous series representation. This means that our quadratic extension should be real and our grossencharacter should not come from a grossencharacter of $\mathbf{Q}$.

To simplify matters we fix L equal to $\mathbf{Q}(\sqrt{2})$. With this choice of discriminant only the prime $p = 2$ ramifies in L. We also want our grossencharacter to be as unramified as possible. Therefore we set

(7.41) $$\lambda((a_{\mathfrak{p}})) = \prod_{j=1}^{2} |a_{\infty_j}|^{it_j} \prod_{\mathfrak{p}\ \text{finite}} |a_{\mathfrak{p}}|_{\mathfrak{p}}^{is_{\mathfrak{p}}}$$

where $\infty_j$ denotes the place at infinity corresponding to the embedding

$$a + b\sqrt{2} \to a + (-1)^{j+1} b\sqrt{2} .$$

(The constraint to keep in mind is that $\lambda$ must not arise from a grossencharacter of $\mathbf{Q}$.)

Now the function defined by (7.41) is clearly a character of the idele group $A^x$ of L. It will be a grossencharacter of L precisely when

(7.42) $$\lambda(a) = 1$$

for all principal idèles $a$. This last condition imposes certain conditions on $t_1$ and $t_2$ that we shall now describe.

Let $I_S$ denote the set of S units of L with S the set of infinite primes. Then $I_S$ is the group of units in the ring of integers of L and is generated (in this case) by $-1$ and the fundamental unit $\sqrt{2} + 1$. Since (7.42) must hold in particular for these S-units we must have

(7.43) $$\lambda(-1) = 1$$

and

(7.44) $$\lambda(\sqrt{2} + 1) = 1 .$$

This first equation puts no restriction of $t_1$ and $t_2$, but the second implies that

$$|\sqrt{2} + 1|^{it_1} |-\sqrt{2} + 1|^{it_2} = 1$$

or, taking logarithms, that

(7.45) $$it_1 \log|\sqrt{2} + 1| + it_2 |\log \sqrt{2} - 1| = 2k\pi i$$

with $k \in \mathbf{Z}$. We note that infinitely many pairs $\{t_1, t_2\}$ may satisfy (7.45). However, since $(\sqrt{2}+1) = (\sqrt{2}-1)^{-1}$, each pair $t_1, t_2$ must be chosen such that $i(t_2 - t_1)\log(\sqrt{2}-1) = 2k\pi i$, i.e. such that

(7.46) $$s = t_2 - t_1 = \frac{2k\pi}{\log(\sqrt{2}-1)}, \quad k \in \mathbf{Z}.$$

*Henceforth we fix* $k = 1$ *in (7.46) and we assume* $t_1 + t_2 = 0$. Having chosen fixed $t_1, t_2$ satisfying (7.46) there will be at least one choice of $\{s_{\mathfrak{p}}\}$ such that the formula (7.41) defines a grossencharacter of L. In fact since the class number of $\mathbf{Q}(\sqrt{2})$ is 1 there will be exactly one choice of such $\{s_{\mathfrak{p}}\}$. (Cf. the discussion in [Tate], p. 343.) Note that the resulting grossencharacter $\lambda$ can not possibly be of the form $\mu \circ N$ with $\mu$ a grossencharacter of $\mathbf{Q}$ since our choice of k in (7.46) implies $t_1 \neq t_2$.

From this discussion we conclude that the representation

(7.47) $$\pi(\lambda) = \pi_{it_1, it_2} \otimes \Big( \bigotimes_{p<\infty} \pi_p \Big)$$

occurs in $R_0^{\psi}$ if $\psi$ denotes the central character of $\pi(\lambda)$ and $\pi_p$ is the representation of $GL(2, \mathbf{Q}_p)$ associated to $\lambda_{\mathfrak{p}}$ (or the pair $\lambda_{\mathfrak{p}}, \lambda_{\mathfrak{p}'}$ if both $\mathfrak{p}$ and $\mathfrak{p}'$ divide p). Note that $\pi_{it_1, it_2}$ is trivial on the center of $GL(2, \mathbf{R})$ since $t_1 + t_2 = 0$. Thus finally we can prove:

THEOREM 2.11. *The continuous series representation* $\pi_{is}$, with

$$s = \frac{2\pi}{\log(\sqrt{2}-1)},$$

*occurs discretely in* $L^2(\Gamma_0'(2)\backslash SL(2, \mathbf{R}))$.

*Proof.* We claim that the conductor of $\pi(\lambda)$ is 2. Indeed since only $p = 2$ ramifies in L, and $\lambda_{\mathfrak{p}}$ is unramified for *every* $\mathfrak{p}$, $\pi_p$ is class 1 for all odd p (by the proof of Proposition 7.8). Thus the conductor of $\pi(\lambda)$ is the conductor of $\pi_2$.

To compute the conductor of $\pi_2$ let $\delta_2$ denote the character $|\ |^{is_{\mathfrak{p}}}$ where $\mathfrak{p} = \sqrt{2}$ lies above 2. Let $\omega_2$ denote the character (of order 2) of $\mathbf{Q}_2^{\times}$ attached to the quadratic extension $\mathbf{Q}_2(\sqrt{2})$. Then $\pi_2 = \pi(\delta_2, \delta_2\omega_2)$.

(This follows from the fact that $|\alpha_{\mathfrak{p}}|_{\mathfrak{p}}^{is_{\mathfrak{p}}} = |N(\alpha_{\mathfrak{p}})|_{Q_2}^{is_{\mathfrak{p}}}$ if $\mathfrak{p} = \sqrt{2}$.) So since $\omega_2$ has conductor 2 (or $20_2$) our claim is an immediate consequence of (4.20).

To prove the theorem let $H(t_1, t_2)$ denote the subspace of the representation space of $\pi(\lambda)$ in $L_0^2(G_Q \backslash G_A, \psi)$ consisting of functions which transform according to $\psi_2$ under the action of $K_0^2$. By our claim above $H(t_1, t_2)$ is equivalent (as a $GL(2, R)$-module) to $\pi_{it_1, it_2}$. But the restriction of $\pi_{it_1, it_2}$ to $SL(2, R)$ is irreducible and equivalent to $\pi_{is}$ where $s = t_1 - t_2$. (Cf. the "Notes and References" in Section 4.) Therefore, since

$$Z_\infty^+ G_Q \backslash G_A / K_0^N \cong \Gamma_0(2) \backslash SL(2, R) ,$$

$H(t_1, t_2)$ is equivalent *as an* $SL(2, R)$-*module* to an irreducible subspace of $L^2(\Gamma_0(2) \backslash SL(2, R))$ belonging to $\pi_{is}$. □

### D. Connections with Class Field Theory

Here we shall be vague as well as brief. Our remarks are directed toward the reader who would like to have *some* idea of the relationship between Jacquet-Langlands' theory and (possible future developments in) class field theory.

Let $F$ denote a global field and $L$ a Galois extension of $F$.

The central concern of class field theory is then the arithmetic study of such extensions. In particular one should obtain a description of how the F-primes split in $L$ and a "reciprocity law" describing the Galois group of this extension in terms of $F$.

A Galois extension is said to be *abelian* if its Galois group is abelian. In the case of abelian extensions the fundamental reciprocity law may be described as follows. Let $\mathfrak{G} = G(L/F)$ denote the Galois group of the abelian extension $L/F$ and let $\chi$ denote a character of $\mathfrak{G}$. Then one can canonically associate to $\chi$ a character of the group $I_F(d_F)$ (which is again denoted by $\chi$). (Here $I_F(d_F)$ is the group of F-ideals relatively prime to the discriminant $d_F$ of $F$; if $v$ is an L-*unramified* finite prime of F and w is a fixed prime of L dividing v one can set $\chi(v)$ equal to $\chi$ evaluated at the Frobenius automorphism at v and extend $\chi$ to $I_F(d_F)$ by multiplicativity.) The "*abelian*" L-*function* corresponding to $\chi$ is

$$(7.48) \qquad L(s,\chi;L/F) = \prod_{v \text{ unramified}} (1-\chi(v)(Nv)^{-s})^{-1}$$

$$= \sum_{\mathfrak{A}} \chi(\mathfrak{A})\, N(\mathfrak{A})^{-s}$$

the summation extending over all ideals in $I_F(d_F)$.

On the other hand suppose $\psi$ is a grossencharacter of F, i.e. a unitary character of $GL(1,A(F)) = A(F)^{\times}$ trivial on $GL(1,F) = F^{\times}$. Suppose also that $\psi = \Pi\psi_v$ and that $\psi_v$ is unramified for each v not dividing $d_F$. This $\psi$ defines a "Hecke character" of $I_F(d_F)$ through the formula

$$(7.49) \qquad \chi(v) = \psi_v(\tilde{\omega}_v)$$

and the fundamental reciprocity law of abelian class field theory may be formulated as the assertion that *every character* $\chi$ *of* $\mathfrak{G}$ *(viewed as a character of* $I_F(d_F)$*) arises in this way. That is, every character* $\chi$ *of* $\mathfrak{G}$ *is associated to some grossencharacter* $\psi$ *of* F.

Now each grossencharacter $\psi$ of F has associated to it a Hecke L-function $L(s,\psi)$. Therefore the reciprocity law above may alternately be written in the form

$$(7.50) \qquad L(s,\chi;L/K) = L(s,\psi)$$

where $\psi$ is some appropriate grossencharacter of F. Note that the left side of (7.50) is an Euler product whose v-th factor depends on how v splits in L. But the right side of (7.50) is an Euler product whose v-factor depends only on $\psi_v$. Thus the arithmetic significance of an equality such as (7.50) is evident.

Summing up, the fundamental result of *abelian* class field theory asserts the following. Suppose $\chi$ is any character of the Galois group of an extension L/F. Then *there is a canonical correspondence which associates to* $\chi$ *an automorphic form* $\psi(\chi)$ *on* GL(1) *(over* F*) such that*

$$L(s,\chi;L/K) = L(s,\psi(\chi)) \ . \tag{7.51}$$

This result may also be expressed without explicit reference to the extension field L by considering in place of L the *maximal* abelian extension $F_{ab}$ of F. If $\mathfrak{G}_{ab}$ denotes the Galois group of $F_{ab}/F$ one may assert that the dual of $\mathfrak{G}_{ab}$ is isomorphic to the torsion subgroup of the dual of the idele class group of F.

Now if we want to consider *not* necessarily abelian extensions of F it is natural to introduce the Galois group $\mathfrak{G}$ of the *algebraic closure* of F over F. A fundamental problem of non-abelian class field theory then is to describe the set $\mathfrak{G}^*(n)$ consisting of equivalence classes of irreducible unitary representations of $\mathfrak{G}$ of dimension n. Note that $\mathfrak{G}^*(n)$ is simply the dual group of $\mathfrak{G}_{ab}$ in case $n = 1$. (This is *not* to say that $\mathfrak{G} = \mathfrak{G}_{ab}$; rather every character of $\mathfrak{G}$ is trivial on the commutator subgroup of $\mathfrak{G}$ and hence defines a character of $\mathfrak{G}_{ab}$.) In case $n = 1$ the solution to the problem is therefore given by abelian class field theory.

If $n = 2$, let $\hat{X}$ denote the collection of automorphic cusp forms on GL(2) over F, i.e. irreducible unitary representations $\pi$ of GL(2,A(F)) which occur in $R_0^{\psi}$ for some appropriate $\psi$. Then what one would hope to be able to do is define an injection from $\mathfrak{G}^*(2)$ into X and describe its image completely. (This is analogous to specifying the torsion subgroup of the dual of $F^x\backslash A(F)^x$ in case $n = 1$.)

One general result along these lines is due to Jacquet-Langlands. Before describing it let us point out that Theorem 7.11 of Subsection B is more or less equivalent to the following. Suppose $\pi'$ is a monomial element of $\mathfrak{G}^*(2)$, i.e. $\pi'$ is a representation induced from a suitable character $\lambda$ of a subgroup of $\mathfrak{G}$ of index 2. Then one may associate to $\pi'$ an automorphic cusp form $\pi(\pi')$ on GL(2). This is so because the *Artin L-function* $L(\pi',s)$ is (in this case) entire and coincides with the L-function $L(s,\pi(\pi'))$. (Note that $\lambda$ may be identified with a character of the idele group of a quadratic extension of F, whence Theorem 7.11.

The result of Jacquet-Langlands alluded to above represents a vast generalization of Theorem 7.11 and to state it properly requires additional terminology and reformulation.

If $F$ is a local field let $C_F$ denote the multiplicative group of $F$, and if $F$ is a global field, let $C_F$ denote the idele class group of $F$. If $L$ is a finite Galois extension of $F$ then the *Weil group* $W(L/F)$ is a certain extension of $\mathfrak{G} = G(L/F)$ by $C_L$. In particular $W(L/F) = C_F$ if $L = F$ and $W(L/F)/C_L = G(L/K)$ in general. If $F$ is global, $v$ is a place of $F$, and $w$ in $L$ divides $v$, there is a uniquely determined homomorphism $\alpha_v$ of $W(L_w/F_v)$ into $W(L/F)$ (uniquely determined up to inner automorphism by an element of $C_L$).

If $F$ is a local field then to every representation $\sigma$ of $W(L/F)$ one can associate a local "L-function" $L(s,\sigma)$ and local factor $\varepsilon(s,\sigma)$ both of which depend only on the class of $\sigma$. If $F$ is global and $\sigma$ is a semi-simple (complex) representation of $W(L/F)$ then $\sigma_v = \sigma \circ \alpha_v$ is a representation of $W(L_w/F_v)$ whose class is determined by that of $\sigma$. Therefore we set

$$L(s,\sigma) = \prod_v L(s,\sigma_v) \; . \tag{7.52}$$

The function defined by this product (in a certain half-plane) extends to a meromorphic function in all of $\mathbb{C}$ and satisfies the functional equation

$$L(s,\sigma) = \varepsilon(s,\sigma) L(1-s,\tilde{\sigma}) \tag{7.53}$$

where $\tilde{\sigma}$ is contragredient to $\sigma$ and $\varepsilon(s,\sigma) = \prod_v \varepsilon(s,\sigma_v)$.

Now suppose $\sigma$ is a *two* dimensional representation of $W(L/F)$. From Theorem 6.14 it follows that there is *at most* one irreducible admissible representation $\pi$ of $\mathcal{H}(G = GL(2,F))$ such that $\pi\left(\begin{bmatrix} a & 0 \\ 0 & a \end{bmatrix}\right) = \det(\sigma(a)) I$ and such that for all characters $\chi$ of $F^\times = C_F$,

$$L(s,\chi \otimes \pi) = L(s,\chi \otimes \sigma)$$

$$L(s,\chi^{-1} \otimes \tilde{\pi}) = L(s,\chi^{-1} \otimes \tilde{\sigma})$$

and

$$\varepsilon(s, \chi \otimes \pi) = \varepsilon(s, \chi \otimes \sigma) .$$

(Here $\chi$ lifts to a one-dimensional representation of $W(L/F)$ through the natural homomorphism from $W(L/F)$ onto $W(F/F) = C_F$.) We note that $\pi = \pi(\lambda)$ if $L$ is a separable quadratic extension of $F$ and $\sigma$ is the representation of $W(L/F)$ induced from the character $\lambda$ of the subgroup $C_L = W(L/L)$. If $\sigma$ is the direct sum of one dimensional representations of $W(L/F)$ lifted (as above) from the characters $\mu_1$ and $\mu_2$ of $F^{\times}$ then $\pi(\sigma) = \pi(\mu_1, \mu_2)$. In general, $\pi$, *if it exists at all*, will be denoted by $\pi(\sigma)$.

THEOREM 7.12. *Suppose $F$ is a global field and $\sigma$ is a two-dimensional semi-simple representation of the Weil group of the separable quadratic extension $L$ of $F$. Let $\psi = \det(\sigma)$. Suppose also that for every character $\chi$ of $C_F$ both $L(s, \chi \otimes \sigma)$ and $L(s, \chi^{-1} \otimes \tilde{\sigma})$ are entire functions of $s$ bounded in vertical strips of finite width. Then $\pi(\sigma_v)$ exists for every place $v$ and $\pi(\sigma) = \otimes \pi(\sigma_v)$ is a representation of $GL(2, A(F))$ which occurs in $R_0^{\psi}$.*

Observe that by the above remarks this result includes Theorem 7.11 as a special case. (Take $\sigma$ equal to the representation of the global Weil group induced from the character $\lambda$ of $W(L/L)$. Then $\sigma_v = \mathrm{Ind}(\lambda_w)$ if $v$ does *not* split in $L$ but lies under $w$, and is the direct sum of two characters if $v$ *does* split in $L$.) In this case $L(s, \sigma)$ is known to be entire by the abelian theory. On the other hand, for *function* fields $F$ and *arbitrary* representations $\sigma$, the L-functions $L(s, \chi \otimes \sigma)$ are also known to be entire. Therefore Theorem 7.12 is a "real" Theorem in this case as well.

For arbitrary fields it should be remarked that the converse to Theorem 7.12 follows immediately from the "only if" part of Theorem 6.18 since

$$L(s, \sigma) = L(s, \pi(\sigma)) . \tag{7.54}$$

That is, *if* $\pi(\sigma_v)$ can be shown to exist for every place v, and *if* $\bigotimes_v \pi(\sigma_v)$ can be shown to be a cusp form, then $L(s,\sigma)$ *will* be entire! Although this hardly constitutes a feasible attack on the conjecture that $L(s,\sigma)$ is always entire it does demonstrate (once again) the interest of Theorem 6.18.

Finally we observe that the (7.54) is a natural generalization of the abelian reciprocity law (7.51).

## FURTHER NOTES AND REFERENCES

The application of Weil's representation to the construction of irreducible representations of other semi-simple groups is the subject matter of [Gelbart 4] and [Gross-Kunze]. The example of Subsection C is taken from [Gelbart 2].

Several important conjectures and methods related to Langlands' general "philosophy of L-functions" are to be found in [Langlands 2]. These conjectures are motivated by the results of [Langlands 3] and [Jacquet-Langlands] represents an attempt to verify them for GL(2).

If one forgets about L-functions (for the moment) it still seems to be of interest to formulate the following:

*Problem* A. Given two groups G′ and G, what natural relations are there (if any) between automorphic forms on G′ and G?

Here we are assuming that G and G′ are such that the notion of automorphic form on G or G′ is meaningful. If G is an algebraic group defined over a global field F then an automorphic form must be an irreducible representation of the adele group of G which occurs in the obvious $L^2$ space. An important example is when $G = GL(2)$ (over F) and G′ is the multiplicative group of quadratic extension of F. In this case a solution to Problem A is provided by Theorem 7.11 (and this solution is natural because the appropriate L-functions coincide).

In general one would hope that a solution to Problem A would make it possible to construct automorphic forms on certain groups using automorphic forms on simpler groups and to obtain arithmetic information from the explicit relations between these forms. Such is the case when $G = GL(2)$ and $G'$ is the multiplicative group of a quaternion algebra (see Section 10). Other examples include those which appear in [Jacquet] (where $G' = GL(2)$ over $F$ and $G = GL(2)$ over a *quadratic extension* of $F$) [Gelbart 3] (where $G'$ is (an extension of) Weil's metaplectic group and $G = GL(2)$), and unpublished work of Shintani.

Our discussion in Subsection D is adapted from [Shalika], [Langlands 2] and [Jacquet-Langlands]. The construction (and the motivation for the construction) of the groups $W(L|F)$ is explained in [Weil 5]. Weil also discusses the L-functions $L(s, \sigma)$ systematically developed in [Langlands 3]. These L-functions generalize the "non-abelian" L-series introduced in [Artin] as well as the "abelian" L-series of Hecke.

## §8. EISENSTEIN SERIES AND THE CONTINUOUS SPECTRUM

As before let $R^{\psi}$ denote the natural representation of $G_A$ in $L^2(G_Q \backslash G_A)$. The purpose of this section is to isolate the *continuous spectrum* of $R^{\psi}$ from the discrete spectrum. This will be accomplished by constructing an intertwining operator between a certain direct integral of induced representations and the restriction of $R^{\psi}$ to an appropriate invariant subspace whose orthocomplement decomposes discretely. Since this construction involves the theory of Eisenstein series and is not completely straightforward some additional remarks might be in order.

Our motivating theme here is the principle which asserts that a unitary representation can be decomposed by spectrally decomposing the self-adjoint operators which commute with it. In the context of $R^{\psi}(g)$ this means (as in Section 2) that functions in $L^2(G_Q \backslash G_A, \psi)$ must be expanded in terms of $\psi$-automorphic forms on GL(2). Eisenstein series enter the picture because they produce sufficiently many automorphic forms to decompose the "continuous" part of $L^2(G_Q \backslash G_A, \psi)$. Obstacles arise because these forms themselves are not square-integrable and therefore an intermediary Fourier transform is required. The reader might keep in mind that this situation is analogous to what happens when the regular representation of $\mathbb{R}$ is decomposed using the (non square-integrable) "forms" $e^{ixy}$.

We must emphasize that it is *not* our purpose to develop the theory of Eisenstein series from scratch. Rather our purpose is to explain the *role* Eisenstein series play in the decomposition of $R^{\psi}(g)$ and (in Section 9) in the proof of the Selberg trace formula. Therefore several facts from the theory of Eisenstein series will be recalled without proof. For details and complete proofs the reader is referred to [Selberg], [Langlands 4], [Harish-

Chandra 2], [Langlands 5], [Kubota] and "Further Notes and References" at the end of this section.

A. Some Preliminaries

Let $Z_\infty^+$ denote the group

$$\left\{\begin{bmatrix} r & 0 \\ 0 & r \end{bmatrix} \epsilon\, Z_\infty : r > 0\right\}$$

and put

$$X = Z_\infty^+ G_Q \backslash G_A \,. \tag{8.1}$$

Since it is slightly more convenient (but no less general) to deal with $L^2(X)$ in place of $L^2(G_Q\backslash G_A, \psi)$ we shall do so until further notice. Therefore $R(g)$ will henceforth denote the natural (unitary) representation of $G_A$ in $L^2(X)$. In Subsections C and D the theory of Eisenstein series will be used to produce the preliminary decomposition

$$R = R_0^+ \oplus R_1 \tag{8.2}$$

where $R_0^+$ is a *direct* sum of certain (unknown) irreducible representations and $R_1$ is a direct *integral* (or *continuous* sum) of certain (*known*) induced representations $R(g:z)$, $z \,\epsilon\, i\mathbb{R}$.

Throughout this section the following notation will be fixed: B will denote the upper triangular group of G, A the diagonal subgroup, and N the upper niltriangular subgroup. The Haar measure dx will denote the *Tamagawa measure* on $G_A$. This choice of measure is made for normalization purposes only so its precise definition is not important. See [Ono] or [Weil 6] for details.

On any discrete subgroup of $G_A$ we shall use the Haar measure which assigns to any point the measure one, and on any quotient of unimodular groups to which we have assigned Haar measures we shall use the corresponding quotient measure. Thus if we assign to $Z_\infty^+$ the measure which corresponds to Euclidean measure on $(\mathbb{R}_+)^x$ we determine a measure on $G_A^1 = Z_\infty^+\backslash G_A$ and hence also a measure on

$$X = Z_\infty^+ G_Q \backslash G_A$$

whose total volume is *the Tamagawa number of* GL(2).

Similarly, we have uniquely determined measures on $A_A$, $N_A$, $A_\infty^+ = \left\{\begin{bmatrix} a_1 & 0 \\ 0 & a_2 \end{bmatrix} \epsilon A_\infty : a_i > 0\right\}$, and

$$\text{(8.3)} \qquad A_A^1 = \left\{\begin{bmatrix} a_1 & 0 \\ 0 & a_2 \end{bmatrix} \epsilon A_A : |a_1| = |a_2| = 1\right\} = A_\infty^+ \backslash A_A .$$

On $B_A$, which is non-unimodular, we define a left Haar measure by

$$\text{(8.4)} \qquad \int_{B_A} \phi(b)\, d_\ell b = \int_{A_A} \int_{N_A} \phi(an)\, da\, dn$$

and right Haar measure by

$$\text{(8.5)} \qquad \int_{B_A} \phi(b)\, d_r b = \int_{N_A} \int_{A_A} \phi(na)\, dn\, da .$$

We recall that

$$\text{(8.6)} \qquad d_r b = \delta_B(b)\, d_\ell b$$

where

$$\delta_B(b) = \left|\frac{a_1}{a_2}\right|, \quad \text{if} \quad b = \begin{bmatrix} a_1 & n \\ 0 & a_2 \end{bmatrix} .$$

Also that $d(ana^{-1}) = \delta_B(a)\, dn$ and that

$$\text{(8.7)} \qquad \int_{G_A} f(x)\, dx = c_G \int_{B_A} \int_K f(bk) \delta_B(b)^{-1} d_r b\, dk$$

for some constant $c_G$.

Now fix once and for all an isomorphism

$$t \to h_t = \begin{bmatrix} e^t & 0 \\ 0 & e^{-t} \end{bmatrix}$$

from $\mathbf{R}$ onto a subgroup $T$ of $A_\infty^+$ (so $A_\infty^+$ is the direct product of $Z_\infty^+$ and $T$). Write

$$x = nah_t kz \tag{8.8}$$

where $n \in N_{\mathbf{A}}$, $a \in A_{\mathbf{A}}^1$, $t \in \mathbf{R}$, $k \in K$, and $z \in Z_\infty^+$. Since the number $t$ is uniquely determined by $x$ we may denote it by $H(x)$. Note that

$$H(b) = \log \left|\frac{a_1}{a_2}\right|^{\frac{1}{2}} \quad \text{if} \quad b = \begin{bmatrix} a_1 & 0 \\ 0 & a_2 \end{bmatrix}, \tag{8.9}$$

and

$$\delta_B(b) = e^{2H(b)} \quad \text{if} \quad b \in B_{\mathbf{A}} . \tag{8.10}$$

The integration formula corresponding to the decomposition (8.8) is then

$$\int_{Z_\infty^+\backslash G_{\mathbf{A}}} f(x)\,dx = c_G \int_{N_{\mathbf{A}}} \int_{A_{\mathbf{A}}^1} \int_{-\infty}^{\infty} \int_K f(nah_t k) e^{-2t} dn\,da\,dt\,dk . \tag{8.11}$$

Finally we recall the notion of *Siegel domain* appropriate to this context. For $c > 0$ put

$$\gamma(c) = \omega A_\infty^+(c) K \tag{8.12}$$

where $\omega$ is some relatively compact subset of $N_{\mathbf{A}} A_{\mathbf{A}}^1$ and

$$A_\infty(c) = \{a \in A_\infty^+ : H(a) \geq \log c\} . \tag{8.13}$$

The Siegel domain defined by (8.12) may be chosen so that $G_{\mathbf{A}} = G_{\mathbf{Q}}\gamma(c)$.

A function $f(x)$ on $G_{\mathbf{A}}$ will be called *slowly increasing on* $\gamma(c)$ if there are constants $C$ and $N$ such that

$$|f(x)| \leq Ce^{-NH(x)}, \quad x \in \gamma(c)$$

and *rapidly decreasing on* $\gamma(c)$ if for every $N$ there is a constant $C_N$ such that

$$|f(x)| \leq C_N e^{NH(x)}, \quad x \in \gamma(c) \ .$$

If $f$ is any continuous function on $Z_\infty^+ G_Q \backslash G_A$ its *constant term* is the function

$$f^0(x) = \int_{N_Q \backslash N_A} f(nx)\,dn \ . \tag{8.14}$$

Note that $f$ is a *cusp form* if and only if its constant term is zero almost everywhere. If $f$ is an *automorphic form* on $GL(2)$ (in the sense of Definition 3.3, with $\psi$ trivial on $Z_\infty^+$) then $f - f^0$ is rapidly decreasing on any Siegel domain. In particular $f - f^0$ is square integrable on $X$. Consequently every *cuspidal* automorphic form is in $L^2(X)$, as already remarked in Section 3.

B. **Analysis of Certain Induced Representations**

The representations in question are induced from the (non-unimodular) subgroup $N_A A_Q A_\infty^+$ of $B_A$. For each $z \in C$ consider the character

$$b = \begin{bmatrix} a_1 & u \\ 0 & a_2 \end{bmatrix} \to e^{zH(b)} = \left|\frac{a_1}{a_2}\right|^{z/2}$$

of $N_A A_Q A_\infty^+$. The resulting induced representation

$$R(x, z) = \operatorname*{Ind}_{N_A A_Q A_\infty^+ \uparrow G_A} e^{zH(b)}$$

operates by right translation in the space $H(z)$ consisting of complex-valued measurable functions $\phi$ on $G_A$ such that

$$\int_K |\phi(k)|^2\,dk < \infty \tag{8.15}$$

and

$$\phi(nag) = e^{(z+1)H(na)}\phi(g) = \left|\frac{a_1}{a_2}\right|^{1/2 + z/2}\phi(g) \tag{8.16}$$

for all $na \in N_A A_Q A_\infty^+$. Note that $R(x,z)$ is actually a representation of $Z_\infty^+ \backslash G_A$.

LEMMA 8.1. *Let* $H$ *denote the Hilbert space of measurable functions* $\phi$ *on* $N_A A_Q A_\infty^+ \backslash G_A$ *satisfying (8.15) and define a representation* $R'(x,z)$ *in* $H$ *through the formula*

$$R'(x,z)\phi(g) = \phi(gx)e^{(z+1)H(gx)}e^{-(z+1)H(g)} . \tag{8.17}$$

*Then the representations* $R(x,z)$ *and* $R'(x,z)$ *are equivalent. The operator intertwining them is*

$$I(z) : \phi(g) \to \phi(g)e^{-(z+1)H(g)} . \tag{8.18}$$

*Proof.* Completely straightforward since $H(nag) = H(na) + H(g)$. □

LEMMA 8.2. *The Hilbert space adjoint of* $R(x,z)$ *is* $R(x^{-1}, -\bar{z})$. *Consequently* $R(x,z)$ *is unitary whenever* $\mathrm{Re}(z) = 0$.

*Proof.* To prove this lemma it is convenient to use the "multiplier realization" of $R(x,z)$, namely $R'(x,z)$. Indeed since $N_A A_Q A_\infty^+ \backslash G_A$ is compact we can find a real-valued function $\beta$ in $C_c^\infty(G_A)$ such that $\int_{N_A A_Q A_\infty^+} \beta(bx)\, d_r b = 1$ for $x \in G_A$. Consequently

$$(\phi, \psi) = \int_{G_A} e^{2H(x)}\beta(x)\phi(x)\overline{\psi(x)}\,dx \tag{8.19}$$

for $\phi, \psi$ in $H$ (by (8.7) and (8.15)). The desired equality

$$(R'(x,z)\phi, \psi) = (\phi, R'(x^{-1}, -\bar{z})\psi)$$

then follows from (8.19) by an obvious change of variables. □

Our aim is to show that the direct integral

$$\int_{\substack{Re(z)=0 \\ Im(z)>0}} R(x,z)\, d|z| \tag{8.20}$$

(of unitary representations $R(x,z)$) is equivalent to a subrepresentation of (the unitary) representation $R(x)$. We shall not need the representations $R(x,z)$ *with negative imaginary part* since $R(x,-z)$ will turn out to be equivalent to $R(x,z)$.

Before explaining why (8.20) is natural we shall need to further analyze its summands. Indeed the representations $R(x,z)$ are themselves not irreducible and it is of some interest to know how they decompose.

By Mackey's "Inducing in Stages Theorem" $R(x,z)$ may be described by first inducing $e^{zH(na)}$ from $N_{\mathbb{A}}A_{\mathbb{Q}}A_\infty^+$ to $N_{\mathbb{A}}A_{\mathbb{A}}$ and then inducing the resulting representation (call it $T(z)$) from $N_{\mathbb{A}}A_{\mathbb{A}}$ to $G_{\mathbb{A}}$. Now the representation space of $T(z)$ consists of functions on $N_{\mathbb{A}}A_{\mathbb{A}}$ which are (at least) left $N_{\mathbb{A}}$-invariant. Therefore $T(z)$ may be viewed as a representation of $A_{\mathbb{A}}$. In fact if $R_{\mathbb{A}}$ denotes the regular representation of $A_{\mathbb{Q}}\backslash A_{\mathbb{A}}^1$ (a compact abelian group) then $T(z)$ is equivalent to the tensor product of $R_{\mathbb{A}}$ with the one-dimensional representation $e^{(z+1)H(a)}$ of $A_\infty^+$. (Since $H\left(\begin{bmatrix} a_1 & 0 \\ 0 & a_2 \end{bmatrix}\begin{bmatrix} a_1' & 0 \\ 0 & a_2' \end{bmatrix}\right) = H\left(\begin{bmatrix} a_1 & 0 \\ 0 & a_2 \end{bmatrix}\right) + H\left(\begin{bmatrix} a_1' & 0 \\ 0 & a_2' \end{bmatrix}\right)$ the action of $T(z)$ on $L^2(A_{\mathbb{Q}}A_\infty^+\backslash A_{\mathbb{A}})$ is given by $\phi(y) \to \phi(yx)e^{(z+1)H(x)}$.) Consequently the following proposition is immediate.

PROPOSITION 8.3. *Let* $\Lambda$ *denote the set of characters* $\lambda$ *of* $A_{\mathbb{Q}}A_\infty^+\backslash A_{\mathbb{A}}$. *Let* $R(x,\lambda,z)$ *denote the principal series representation of* $G_{\mathbb{A}}$ *induced from the character* $e^{zH(a)}\lambda(a)$ *of* $N_{\mathbb{A}}A_{\mathbb{A}}$. *Then*

$$R(x,z) = \bigoplus_{\lambda\in\Lambda} R(x,\lambda,z) \; . \tag{8.21}$$

The definition below makes it possible to complete our analysis of the space of $R(x, z)$.

DEFINITION 8.4. Suppose $\tau$ is an irreducible representation of K. Then we say that $\phi \in C^{\infty}(N_A A_Q A_{\infty}^{+} \backslash G_A)$ is *of type* $\tau$ if for any $x \in G_A$, $\phi(xk)$ is contained in a subspace of $L^2(K)$ on which the right regular representation of K is equivalent to $\tau$ (i.e. $\phi(xk)$ transforms according to $\tau$).

The space of all such $\phi$ will be denoted by $(\tau)$; this space is finite-dimensional and has a natural inner product given by

$$(\phi_1, \phi_2) = c_G \int_K \phi_1(k)\, \overline{\phi_2(k)}\, dk \ .$$

If L denotes the collection of all such $(\tau)$ then

$$\text{(8.22)} \qquad H = \bigoplus_{(\tau) \in L} (\tau) \ .$$

C. Eisenstein Series

Roughly speaking, Eisenstein series provide a map from the space of $R(g, z)$ to a space of automorphic forms on $G_A$. More precisely this theory provides an isomorphism between the space of (8.20) and the continuous spectrum of $L^2(X)$ which commutes with the natural action of $G_A$. The basic Eisenstein series are defined as follows.

For each $\phi \in (\tau)$, $(\tau) \in L$ (cf. Definition 8.4) consider the function

$$\text{(8.23)} \qquad (I^{-1}(z)\phi)(g) = \phi(x) e^{(z+1)H(g)} \ .$$

Since $I^{-1}(z)$ is the "inverse" of $I(z)$ this function belongs to $H(z)$ (since $\tau \in H$). In particular it transforms according to (8.16) under the action of $N_A A_Q A_{\infty}^{+}$. The *Eisenstein series associated to* $\phi$ is (by definition) the series

$$(8.24)\qquad \begin{aligned} E(g,\phi,z) &= \sum_{\gamma\in B_Q\backslash G_Q} (I^{-1}(z)\phi)(\gamma g) \\ &= \sum_{\gamma\in B_Q\backslash G_Q} \phi(\gamma g)\, e^{(z+1)H(\gamma g)} . \end{aligned}$$

REMARK 8.5. The Eisenstein series just introduced generalizes the "classical" Eisenstein series

$$(8.25)\qquad \begin{aligned} E(x+iy,z) &= \sum_{\gamma\in\Gamma\cap N\backslash\Gamma} f_\mu(\gamma(x+iy)) \\ &= \sum_{\substack{(c,d)=1\\ c,d\in \mathbf{Z}}} \frac{y^{\frac{z+1}{2}}}{|cz+d|^{z+1}} \end{aligned}$$

described in Example 2.5. Indeed $E(g,\phi,z) = E(x+iy,z)$ precisely when $\phi$ is identically one and $g = \begin{bmatrix} y & x\\ 0 & 1\end{bmatrix}$ is real, $y > 0$.

In general it is known from the theory of Selberg-Langlands that

(1) the series defining $E(x,\phi,z)$ converges uniformly for $x$ in compact subsets of $Z_\infty^+\backslash G_A$ and $z$ in compact subsets of

$$(8.26)\qquad D = \{z : \mathrm{Re}(z) > 1\} ;$$

(2) for $z\in D$ this series defines a K-finite eigenfunction of $\mathfrak{z}$ (the center of the universal enveloping algebra of $G_\infty$) which is slowly increasing in all Siegel domains, i.e. $E(x,\phi,z)$ is an automorphic form;

(3) the *constant term* of $E(x,\phi,z)$, i.e.

$$(8.27)\qquad E^0(x,\phi,z) = \int_{N_Q\backslash N_A} E(nx,\phi,z)\,dn$$

is given by the expression

$$(8.28)\qquad \phi(x)\,e^{(z+1)H(x)} + M(z)\phi(x)\,e^{(-z+1)H(x)}$$

where $M(z)$ is a uniquely determined analytic function (depending on $(\tau)$) which maps $D$ into the space of linear operators on $(\tau)$: the adjoint of $M(z)$ is $M(\bar{z})$;

(4) $E(x,\phi,z)$ and $M(z)\phi$ can be analytically continued as meromorphic functions of $z$ *regular on* $i\mathbb{R}$: the only *possible* pole to the right of $i\mathbb{R}$ is simple and occurs at $z=1$, simultaneously for $E(x,\phi,z)$ and $M(z)\phi$;

(5) $E(x,\phi,z)$ is slowly increasing if $\mathrm{Re}(z)=0$; and

(6) the following functional equations are satisfied:

$$M(z)M(-z)\phi = \phi \, , \tag{8.29}$$

and

$$E(x, M(z)\phi, -z) = E(x,\phi,z) \, . \tag{8.30}$$

We note that if $\tau$ is trivial then $M(z)$ is equal to $\xi(z)/\xi(z+1)$ with

$$\xi(z) = \pi^{-z/2}\Gamma(z/2)\zeta(z) \, .$$

Therefore the analytic continuation and functional equation of $M(z)$ is (in this case) quite believable.

The significance of the facts (1)-(6) for the decomposition of $R$ can be explained as follows. When $R(z)=0$, both $R(x,z)$ and $E(x,\phi,z)$ are defined (the latter in the sense of analytic continuation). The representation $R(x,z)$ is unitary by Lemma 8.2, and $E(x,\phi,z)$ is an automorphic form by (2) and (5) above. Our claim is that the map

$$\phi \to E(x,\phi,z) \tag{8.31}$$

"essentially" intertwines the space of $R(x,z)$ with a subrepresentation of $R(x)$.

Indeed a (formal) computation shows that

$$E(x, R(g,z)\phi, z) = R(g)E(x,\phi,z) \tag{8.32}$$

i.e. that the map (8.31) from $\mathrm{H}$ to some space of automorphic forms on $G_{\mathbf{A}}$ commutes with $G_{\mathbf{A}}$. The problem is that $E(x,\phi,z)$ is *not* square integrable and therefore (as remarked earlier) a more intricate Fourier transform is required. Before describing this transform let us directly construct a certain continuous sum of the "principal series" representations $R(x,z)$.

For each $(\tau) \in L$ let $\hat{L}_0^2(E,\tau)$ denote the (one or zero dimensional) space of functions from $\{1\}$ to $(\tau)$ with positive semi-definite inner product

(8.33) $$\frac{1}{2\pi}(a_0(1), \mu(1) b_0(1)) \ .$$

Here $\mu(1)$ is the residue of $2\pi i M(z)$ at $z = 1$ and functions are identified modulo null vectors.

Let $\hat{L}_1^2(E,\tau)$ denote the space of square-integrable functions $a_1(z)$ from $i\mathbf{R}$ to $(\tau)$ such that

(8.34) $$a_1(-z) = M(z) a_1(z)$$

and define an inner product in this space

(8.35) $$(a_1, b_1) = \frac{1}{\pi} \int_{\mathrm{Re}(z)=0} (a_1(z), b_1(z))\, d|z| \ .$$

Note that the Hilbert space

$$\bigoplus_{(\tau)\in L} \hat{L}_1(E,(\tau))$$

is essentially the direct integral of the Hilbert spaces $\mathrm{H}(z)$ with respect to Lebesgue measure on $\{z = iy, y > 0\}$. Finally put

(8.36) $$\hat{L}^2(E,\tau) = \hat{L}_0^2(E,\tau) \oplus \hat{L}_1^2(E,\tau) \ .$$

Now let $\mathrm{H}(\tau)$ denote the space of entire functions with values in $(\tau)$ which are Fourier-Laplace transforms of functions in $C_c^\infty(i\mathbf{R}) \otimes (\tau)$.

For any $a \in H(\tau)$ define

$$\tilde{a}(x) = \frac{1}{2\pi} \int_{Re(z)=z_0} E(x, a(z), z)\, d|z| \tag{8.37}$$

where $z_0$ is arbitrary but to the right of 1. Again from Selberg-Langlands' theory it is known that

(7) $\tilde{a}(x) \in L^2(X)$;

(8) if the closure of $\{\tilde{a} : a \in H(\tau)\}$ in $L^2(X)$ is denoted by $L^2(E,\tau)$ *then*

$$L^2(E) = \bigoplus_{(\tau)\in L} L^2(E,\tau) \tag{8.38}$$

*is precisely the orthocomplement (in* $L^2(X)$) *of the space of cusp forms* $L_0^2(X)$; *and*

(9) for any $a, b$ in $H(\tau)$,

$$\begin{aligned} \int_X \tilde{a}(x)\tilde{b}(x)\,dx &= \frac{1}{2\pi} \int_{Re(z)=z_0} \{(a(z), b(-\bar{z})) + (M(z)a(z), b(\bar{z}))\}\, d|z| \\ &= \frac{1}{2\pi}(\mu(1)\,a(1), b(1)) \\ &\quad + \frac{1}{4\pi}\int_{-i\infty}^{i\infty} (a(z) + M(-z)a(-z), b(z) + M(-z)b(-z)\, d|z| . \end{aligned} \tag{8.39}$$

This last equality results from an application of the residue theorem (which is justified by the fact that the norm of $M(z)$ is bounded at infinity in the strip $\{0 \le Re(z) \le z_0\}$.

We shall now explain the significance of these facts for the decomposition of $R(x)$. Using (7)-(9) one can define an isomorphism $E$ between

$\hat{L}^2(E,\tau)$ and $L^2(E,(\tau))$ (and hence ultimately between $\hat{L}^2(E)$ and a certain privileged subspace of $L^2(X)$) by requiring that E map $(a_0,a_1)$ to $\tilde{a}$ whenever $(a_0,a_1)$ arises from $a \in H(\tau)$ through the formulas

$$a_0(1) = a(1)$$

and

$$a_1(z) = \frac{1}{2}(a(z)+M(-z)\,a(-z)) \quad (z \in i\mathbb{R}) \;. \tag{8.40}$$

Indeed by (9) the correspondence

$$(a_0,a_1) \overset{E}{\longleftrightarrow} \tilde{a}$$

is a linear *isometry* between dense subspaces of $\hat{L}^2(E,\tau)$ and $L^2(E,\tau)$. On the other hand by (8) the orthocomplement of $\oplus L^2(E,(\tau))$ is $L_0^2(X)$ and by Theorem 5.1 this latter space decomposes discretely. Consequently the role Eisenstein series play in the decomposition of $R(x)$ is clear.

D. Description of the Continuous spectrum

Recall that the goal of this section was to construct an invariant subspace of $L^2(X)$ whose orthocomplement exhausts the discrete spectrum of $R(x)$. Accepting the theory of Eisenstein series this goal is near at hand.

Our claim is that properties (7)-(9) of the last subsection infer that the subspace of $L^2(X)$ defined by

$$\bigoplus_{(\tau)\in L} E(\hat{L}_1^2(E,\tau))$$

is this sought after subspace.

To see this, let $E_0$ (respectively $E_1$) denote the restriction of the isomorphism E to $\hat{L}_0^2(E,\tau)$ (resp. $\hat{L}_1^2(E,\tau)$). By taking direct sums over all $(\tau)\in L$ we obtain maps from

$$\hat{L}_i^2(E) = \bigoplus_{(\tau)\in L} \hat{L}_i^2(E,\tau) \quad (i = 1,2)$$

which we again denote by $E_i$. If we denote by $L_i^2(E)$ the image of $E_i$ then by (7) each $L_i^2(E)$ belongs to $L^2(X)$ and by (8) and (9)

$$L^2(X) = L_0^2(X) \oplus L_0^2(E) \oplus L_1^2(E) \ . \tag{8.41}$$

For each compactly supported measurable function f on $Z_\infty^+\backslash G_{\mathbf{A}}$ we now introduce the (bounded) operators

$$R(f,z) = \int_{Z_\infty^+\backslash G_{\mathbf{A}}} f(g)R(g,z)dg \tag{8.42}$$

and

$$R(f) = \int_{Z_\infty^+\backslash G_{\mathbf{A}}} f(g)R(g)dg \ . \tag{8.43}$$

With this notation the intertwining property of Eisenstein series (originally expressed by (8.32)) takes the form

$$R(f)E(x,\phi,z) = E(x,R(f,z)\phi,z) \ . \tag{8.44}$$

This last identity is an identity between meromorphic functions of $z$ and it is meaningful in a rigorous sense as soon as f is bi-K-finite (since $R(f,z)\phi$ belongs to a finite collection of $(\tau)$'s whereas $R(x,z)\phi$ need not).

Now the spectrum of $L_0^2(X)$ is known to be discrete by Theorem 5.1 so to prove our claim it remains to observe that $L_0^2(E)$ too is invariant with discrete spectrum. This claim in turn will follow from the fact that $\hat{L}_0^2(E)$ owes its existence to poles of Eisenstein series to the right of $i\mathbf{R}$ and such poles occur only at $z = 1$.

In fact such poles occur at $z = 1$ only when $\tau(k) = \chi(\det k)$ with $\chi$ a character of $\mathbf{Q}^\times\backslash\mathbf{A}^\times$ whose square is trivial on $\mathbf{R}_+^\times$. The corresponding residue of $E(x,\phi,z)$ is then proportional to the function

$$D_\chi(g) = \chi(\det g) \ .$$

Thus a consequence of the lemma below is that $L_0^2(E)$ is equivalent to the direct sum of the one-dimensional representations $D_\chi(g)I$. This lemma gives an explicit formula for the functions $E_i a_i(x)$ and is crucial to the sequel. We include its proof because the computations involved are so typical of the general theory.

LEMMA 8.6. *Suppose* $(a_0, a_1)$ *is such that* $a_0 \in \hat{L}_0^2(E,\tau)$ *and* $a_1$ *is a function of compact support in* $\hat{L}_1^2(E,\tau)$. *Then for almost all* $x$,

$$E_0 a_0(x) + E_1 a_1(x) = i \operatorname*{Res}_{z=1} E(x, a_0(1), z) \tag{8.45}$$
$$+ \frac{1}{2\pi} \int_{-i\infty}^{i\infty} E(x, a_1(z), z)\, d|z|$$

*Proof.* Let $f_0(x) + f_1(x)$ denote the right side of (8.46). To prove the lemma it will suffice to prove that the function $g(x) = f_0(x) + f_1(x) - E_0 a_0(x) - E_1 a_1(x)$ satisfies the following properties:

(i) it is square-integrable;

(ii) it is orthogonal to $L_0^2(X)$; and

(iii) its constant term vanishes almost everywhere.

Indeed these properties imply $g(x) = 0$ a.e.

For convenience, let $E_0(x, \phi, z)$ denote $2\pi i \operatorname*{Res}_{\zeta=z} E(x, \phi, z)$. Then $E_0$ itself is an automorphic form and consequently $E_0 - E_0^0$ is rapidly decreasing on any Siegel domain $\gamma(c)$. But from (8.27) and (8.28) it follows that $E_0^0(x, a_0(1), 1) = (\mu(1) a_0(1)(x)$. Thus

$$\int_{\gamma(c)} |E_0^0(x, a_0(1)\, 1)|^2 dx = \int_\omega \int_{\log c}^{\infty} \int_K |(\mu(1) a_0(1))\,(nah_t k)|^2 e^{-2t} db\,dt\,dk$$
$$= \left( \int_{\log c}^{\infty} e^{-2t} dt \right) \int_\omega \int_K (\mu(1) a_0)(1)(ak)\, db\,dk < \infty ,$$

i.e. $E_0^0$ is square-integrable. So since $E_0 - E_0^0$ is rapidly decreasing, $E_0(x, a_0(1), 1)$ (and hence $f_0(x)$) is square integrable too.

Now consider the term $f_1(x)$. Since $a_1(z)$ has compact support, and $E(x, \phi, z)$ is an automorphic form,

$$|f_1(x) - f_1^0(x)| \leq \int_{-i\infty}^{i\infty} |E_1(x, a_1(z), z) - E_1^0(x, a_1(z), z)| d|z| \leq C_N e^{-NH(x)}$$

(for all N). Therefore $f_1 - f_1^0$ is square integrable. But from (4.28) and the fact that $M(z)a_1(z) = a_1(-z)$ it is easy to see that

$$f_1^0(x) = \frac{1}{\pi} \int_{-i\infty}^{i\infty} a_1(z)(x) e^{(z+1)H(x)} d|z| \ . \tag{8.46}$$

This function, being the Fourier transform of the rapidly decreasing function $a_1(z)$, is also square integrable, and therefore $f_1(x)$ and $g(x)$ are square integrable, as desired.

To prove (ii) it suffices to deal with $f_0(x) + f_1(x)$ alone since $(E_0 a_0)(x)$ and $(E_1 a_1)(x)$ are already known to be orthogonal to $L_0^2(X)$. So suppose h is an arbitrary *automorphic form* in $L_0^2(X)$. Then h is rapidly decreasing, and for any $\phi \in (\tau)$, $z \in D = \{\mathrm{Re}(z) > 1\}$,

$$\int_X E(x, \phi, z) \overline{h(x)} dx = \int_{Z_\infty^+ G_{\mathbf{Q}} \backslash G_{\mathbf{A}}} \int_{\gamma \in B_{\mathbf{Q}} \backslash G_{\mathbf{Q}}} \phi(\gamma x) e^{(z+1)H(\gamma x)} \overline{h(\gamma x)} dx$$

$$= \int_{Z_\infty^+ B_{\mathbf{Q}} \backslash G_{\mathbf{A}}} (\phi(x) e^{(z+1)H(x)}) \overline{h(x)} dx$$

$$= \int_{A_{\mathbf{Q}} \backslash A_{\mathbf{A}}^1} \int_K \phi(ak) e^{(z+1)H(a)} \left( \int_{N_{\mathbf{Q}} \backslash N_{\mathbf{A}}} h(nak) dn \right) da dk$$

$$= 0 \ ,$$

since h is cuspidal. That $f_1(x)$ and $f_2(x)$ are also orthogonal to h follows by analytic continuation. So since $L_0^2(X)$ has a basis consisting of automorphic forms, the proof of (ii) is complete.

To prove (iii), choose a sequence of functions $\{b^n\}$ in $H(\tau)$ so that $(b_0^n, b_1^n)$ approaches $(a_0, a_1)$ in $\hat{L}^2(E,\tau)$. By the residue theorem, and the relation (4.28), the constant term of $\tilde{b}^n$ is

$$(b^n)^0(x) = \int_{N_Q \backslash N_A} \frac{1}{2\pi} \int_{Re(z)=z_0} E(nx, b^n(z), z)\, d|z|\, dn$$

$$= \frac{1}{2\pi} \int_{Re(z)=z_0} \{b^n(z)(x)e^{(z+1)H(x)} + (M(z)b^n(z))(x)e^{(-z+1)H(x)}\} d|z|$$

$$= \frac{1}{2\pi} (\mu(1)b_0^n(1))(x) + \frac{1}{\pi} \int_{Re(z)=0} (b_1^n(z))(x)e^{(z+1)H(x)} d|z| \; .$$

But from (8.46) it is then clear that by a suitable choice of $\{b^n\}$ this last expression can be made to approach $f_0^0(x) + f_1^0(x)$ for all x, as n approaches $\infty$. Therefore the constant terms of $\{E_0 a_0 + E_1 a_1\}$ and $\{f_0 + f_1\}$ are equal almost everywhere, since $\tilde{b}^n$ approaches $E_0 a_0 + E_1 a_1$ in the mean. □

REMARK 8.7. From the crucial identity (8.44) and the identity (8.30) (the functional equation for $E(x, \phi, z)$) it follows that

$$R(f, z)M(z) = M(z)R(f, -z) \; , \tag{8.47}$$

i.e. the operator intertwining the equivalent representations $R(x, z)$ and $R(x, -z)$ is precisely $M(z)$ (summed over all $(\tau) \in L$).

An immediate consequence of Lemma 8.6 and the identity (8.47) is the invariance of the subspace $L_1^2(E)$ and its equivalence (via $E_1$) to the direct integral (8.20).

## E. Summing Up

THEOREM 8.8. *Let* $R_0$ *denote the restriction of* $R$ *to the space of cusp forms and* $R^+$ *the restriction to* $\bigoplus_{\chi} \mathbb{C}D_\chi$. *(Recall that* $\chi$ *is any grossencharacter of* $\mathbb{Q}$ *whose square is trivial on* $(\mathbb{R}_+)^x$ *and* $D_\chi(g) = \chi(\det g)$.*) Then*

$$R = R_0^+ \oplus R_1$$

*where*

$$R_0^+(x) = R_0(x) \oplus R^+(x)$$

*and*

$$R_1(x) = \int_0^\infty R(x, it)\,dt\ .$$

*Furthermore the projection of* $L^2(X)$ *onto the space of* $R_1$ *is* $E_1 E_1^*$.

COROLLARY 8.9. *Theorem 8.8 describes the continuous spectrum of* $L^2(\Gamma\backslash SL(1, \mathbb{R}))$ *for congruence subgroups of* $SL(2, \mathbb{Z})$.

EXAMPLE 8.10 (Theorem 2.6). Let $H(s)$ denote the representation space of the principal series representation of $SL(2, \mathbb{R})$ induced from the character

$$\begin{bmatrix} a & u \\ 0 & a^{-1} \end{bmatrix} \to |a|^{is}\ .$$

Then if $\Gamma = SL(2, \mathbb{Z})$,

$$L^2(\Gamma\backslash SL(2, \mathbb{R})) = L_0^2(\Gamma\backslash SL(2, \mathbb{R})) \oplus \mathbb{C} \oplus \int_0^\infty H(s)\,ds$$

where $\mathbb{C}$ denotes the subspace of $L^2$ spanned by the constant functions.

*Proof.* As an $SL(2, \mathbb{R})$-module, $L^2(\Gamma\backslash SL(2, \mathbb{R}))$ is isomorphic to the subspace of $L^2(Z_\infty^+ G_{\mathbb{Q}}\backslash G_{\mathbb{A}})$ *consisting of right* $K_0$*-invariant functions.* So by

Proposition 8.3 (and the definition of $D_\chi(g)$) this example is an immediate consequence of Theorem 8.8. □

We close this section with a lemma which will be used in Section 9.

LEMMA 8.11. *Suppose* $(\tau) \in L$ *and* $\phi \in (\tau)$. *Then for any* $h \in C_c^\infty(Z_\infty^+ G_Q \backslash G_A)$,

$$(E_1^* h(z), \phi) = \frac{1}{2} \int_X h(x) \overline{E(x, \phi, z)}\, dx$$

for almost all $z \in i\mathbb{R}$.

*Proof.* Suppose $a_1(z)$ is an arbitrary continuous compactly supported function in $\hat{L}_1^2(E, \tau)$. Then

$$\frac{1}{\pi} \int_{-i\infty}^{i\infty} ((E_1^* h)(z), a_1(z))\, d|z| = \int_X h(x) \overline{E_1 a_1(x)}\, dx$$

since both sides equal the inner product of $E_1^* h$ and $a_1$ (the left side by definition of the inner product in $\hat{L}_1^2(E, \tau)$, the right side by definition of the adjoint of $E_1$). Interchanging the order of integration on the right side (justified since $a_1(z)$ is compactly supported) yield the expression

$$\frac{1}{2\pi} \int_{-i\infty}^{i\infty} \left( \int_X h(x) \overline{E(x, a_1(z), z)}\, dx \right) d|z| \; .$$

So by the arbitrariness of $a_1$ the desired conclusion follows. □

## FURTHER NOTES AND REFERENCES

The results from the theory of Eisenstein series sketched in this section are due to [Selberg]. Unfortunately Selberg published them without giving complete proofs. The corresponding results for general semi-simple algebraic groups were proved by Langlands and recorded (with complete details) in [Langlands 4,5]. Similar (but weaker) results were independently obtained in [Gelfand-Graev-Pyatetskii-Shapiro].

In addition to the references already given in the text the reader should consult [Harish-Chandra 3]. There the theory of Eisenstein series is sketched for reductive groups defined over a *finite* field.

Our discussion in this section is adapted from [Arthur]. Similar summaries are to be found in [Duflo-Labesse] and [Godement 2,3]. We have worked over the rationals for the sake of simplicity but it goes without saying that the theory is valid for arbitrary global fields as well.

We note finally that the operator $M(z)$ of (8.47) can be written as a product of local operators $M_p(z)$. Then the operator $M_\infty(z)$ essentially coincides with the intertwining operator of [Kunze-Stein] and [Knapp-Stein].

## §9. THE TRACE FORMULA FOR GL(2)

In this section we give a complete treatment of the Selberg trace formula for GL(2). Our proof is adapted from [Jacquet-Langlands] and [Arthur] and since it is entirely functional analytic it differs substantially from the proof found in [Duflo-Labesse].

In Section 10 the trace formula will be applied to the study of automorphic forms. For further applications (as well as some historical background) the reader is referred to the "Notes and References" at the end of this section and Section 10.

### A. Motivation

Roughly speaking, the Selberg trace formula is an analogue for GL(2) of certain features of abelian harmonic analysis. In representation theoretic terms these may be explained as follows.

#### 1. *The Real Situation*

Let $G$ denote (for the moment only) the additive group of $\mathbf{R}$. Let $\Gamma$ denote the discrete subgroup $\mathbf{Z}$ and $R$ the regular representation of $G$ in $L^2(\Gamma\backslash G)$.

For $f \in C_c^\infty(G)$ define

$$R(f)h(y) = \int_G f(x)(R(x)h)(y)dx$$

so that for $h \in L^2(\Gamma\backslash G)$

$$(R(f)h)(x) = \int_G f(y)h(y+x)dy$$

$$= \int_G f(-x+y)h(y)dy \;.$$

Then

$$R(f)h(x) = \int_{\Gamma\backslash G} K(x,y)h(y)dy \;, \tag{9.1}$$

where

$$K(x,y) = \sum_{\gamma\in\Gamma} f(-x+\gamma+y) \;. \tag{9.2}$$

Since $\Gamma\backslash G$ is compact, the fact that f has compact support implies that this last sum is finite. Hence the kernel $K(x,y)$ is smooth, the integral operator $R(f)$ is of trace class, and

$$\mathrm{tr}\, R(f) = \int_{\Gamma\backslash G} K(x,x)dx = \sum_{\gamma\in\Gamma} f(\gamma) \tag{9.3}$$

(at least when f is of the form $f_1 * f_2$).

A "first form" of the Selberg trace formula for GL(2) amounts to an analogue of (9.3) with $G = GL(2,\mathbb{A})$ and $\Gamma = GL(2,\mathbb{Q})$. This formula is difficult to establish because $\Gamma\backslash G$ is no longer compact. However the *restriction* of R(f) to the *discrete* spectrum of R *does* still possess a trace and an analogue of (9.3) for GL(2) *is* possible.

By analogy with the abelian case the hope is that an explicit trace formula of the type just described could be used to further analyze (the decomposition of) R.

Indeed if $G = \mathbb{R}$ the Poisson summation formula tells us that

$$\sum_{\gamma\in\Gamma} f(\gamma) = \sum_{k\in\mathbf{Z}} \hat{f}(k)$$

thus that

$$\mathrm{tr}\, R(f) = \sum \hat{f}(k) \;. \tag{9.4}$$

On the other hand, we also have

$$\mathrm{tr}\, R(f) = \sum m_j \hat{f}(j) \tag{9.5}$$

if $R = \bigoplus_j m_j R_j$, and $R_j$ denotes the unitary representation $e^{ijx}I$ of G occurring in R with multiplicity $m_j$.

Now equating (9.4) and (9.5) yields the formula

$$\sum m_j \hat{f}(j) = \sum \hat{f}(k) = \operatorname{tr} R(f) \ . \tag{9.6}$$

This formula reflects the fact that each $R_j$ occurs exactly *once* in R. Any analogue of it in a non-compact setting would be highly desirable and for GL(2) would constitute a "second" (and much more explicit) form of the trace formula.

## 2. *The Case of Compact Quotient*

Suppose G is a semi-simple Lie group and $\Gamma$ a discrete subgroup of G such that the quotient space

$$X = \Gamma \backslash G$$

is *compact*. Let R denote the unitary representation of G in $L^2(X)$ given by right translation. Then if f(g) is any continuous compactly supported function on G it is easy to show that the operator $R(f) = \int f(g) R(g) dg$ is compact. Indeed a trivial computation shows that R(f) is an integral operator in (the *compact* space) X with kernel

$$K(x, y) = \sum_{\gamma \epsilon \Gamma} f(x^{-1} \gamma y) \ . \tag{9.7}$$

(Cf. (9.1) and (9.2).) Consequently (by Lemma 5.3) we may write

$$R(g) = \oplus\, m_j R_j \tag{9.8}$$

where $m_j < \infty$ (and $R_j$ runs through the equivalence classes of unitary representations of G which occur in R). To obtain further information concerning this decomposition we obviously need some analogues of (9.3)-(9.6).

Let $\chi_j(g)$ denote the *character* of $R_j(g)$. Then $\chi_j(g)$ is the locally integrable central function on $G$ with the property that

$$(9.9) \qquad \operatorname{tr}(R_j(f)) = \int_G f(g)\chi_j(g)\,dg .$$

The existence of such functions (and in particular the existence of a trace for $R_j(f)$) is a consequence of the general theory of Harish-Chandra. From (9.8) it follows that

$$(9.10) \qquad \sum m_j\,\hat{f}(j) = \operatorname{tr} R(f)$$

where

$$f(j) = \int_G f(g)\chi_j(g)\,dg$$

(cf. (9.5)).

On the other hand, from the explicit realization of $R(f)$ as an integral operator with kernel (9.7), it follows that

$$(9.11) \qquad \operatorname{tr} R(f) = \int_{\Gamma\backslash G} \sum_{\gamma\in\Gamma} f(x^{-1}\gamma x)\,dx$$

(at least when $f = f_1 * f_2$; cf. (9.3)).

So equating (9.10) and (9.11) we have

$$(9.12) \qquad \int_{\Gamma\backslash G} \sum f(x^{-1}\gamma x)\,dx = \operatorname{tr}(R(f)) = \sum m_j \hat{f}(j) .$$

For pedagogical purposes we shall refer to (9.11) as (the first form of) the Selberg trace formula. Some elementary manipulations yield the result

$$(9.13) \qquad \operatorname{tr} R(f) = \sum_{\{\gamma\}} \operatorname{vol}(\Gamma(\gamma)\backslash G(\gamma)) \int_{G(\gamma)\backslash G} f(x^{-1}\gamma x)\,dx$$

where $\Gamma(\gamma)$ and $G(\gamma)$ denote the centralizers of $\gamma$ in $\Gamma$ and G and $\{\gamma\}$ denotes a set of representatives of classes of $\gamma$ in $\Gamma$. The "trace formula" we shall establish below will be an analogue of this last formula. However our interest will be in the non-compact case $G = GL(2, \mathbb{A})$, $\Gamma = GL(2, \mathbb{Q})$.

Because $\Gamma\backslash G$ *is* compact in (9.12) the formula (9.13) makes it feasible to explicitly determine (some of) the multiplicities $m_j$. Indeed the right side of (9.13) is a sum of *distributions*

$$I_\gamma : f \to \mathrm{vol}(\cdot) \int_{G(\gamma)\backslash G} f(x^{-1}\gamma x)\,dx \tag{9.14}$$

which are *invariant* in the sense that

$$I_\gamma(f^y) = I_\gamma(f)$$

if $f^y(x) = f(y^{-1}xy)$. So let $\{\chi\}$ denote the collection of characters of *all* the irreducible unitary representations of G. Each $\chi$ defines a basic invariant distribution on G through the formula

$$f \to \chi(f) = \int_G \chi(g) f(g)\,dg$$

and one expects to be able to find (for each $\gamma$) a measure $m_\gamma(\chi)$ on $\{\chi\}$ such that

$$I_\gamma(f) = \int_{\{\chi\}} \chi(f)\,dm_\gamma(\chi)\ . \tag{9.15}$$

Assuming this *is* possible one obtains from (9.13) and (9.12) the formula

$$\sum m_j\chi_j(f) = \sum_{\{\gamma\}} \int_{\{\chi\}} \chi(f)\,dm_\gamma(\chi)\ . \tag{9.16}$$

The formula (9.16) might be called the "second" or "explicit" form of the trace formula. If its usefulness is not already apparent let us indicate at least one immediate application. Suppose $\chi_\ell$ is the character of a *discrete series representation* of G, i.e. the representation $R_\ell$ occurs discretely in $L^2(G)$, or what amounts to the same thing, the matrix coefficients

$$f_\ell(g) = (R_\ell(g)u, v)$$

are square-integrable on G. Suppose also that $f_\ell(g)$ is actually *integrable* as well as square integrable. (For $SL(2, \mathbb{R})$, this is automatic for all but finitely many discrete series representations.) Then (9.16) is still valid with $f_\ell$ in place of f. But $\chi_j(f_\ell) = 0$ if $\ell \neq j$. So making this replacement in 9.16 one obtains the identity

$$m_j = \sum_{\{\gamma\}} \int_{\{\chi\}} \chi(f)\, dm_\gamma(\chi) \tag{9.17}$$

i.e. one obtains an *explicit* formula for the multiplicity of $R_\ell$ in R (hence also a formula for the dimension of a certain space of automorphic forms by a generalization of the "duality theorem" of Section 2).

As already indicated, the analogue of (9.12) is non-trivial if $\Gamma\backslash G$ is *not* compact and it is this "first" form of the trace formula that will henceforth concern us for GL(2).

3. *The Situation for* GL(2)

As in Section 8, $Z_\infty^+$ will denote the group

$$\{z \in Z_\infty : z = \begin{bmatrix} r & 0 \\ 0 & r \end{bmatrix},\ r > 0\},$$

X the homogeneous space $Z_\infty^+ G_Q \backslash G_A$, and R the unitary representation of $G_A$ in $L^2(X)$ given by

$$R(x)h(y) = h(yx).$$

In Section 8 we explained how the theory of Eisenstein series yields the decomposition

$$R = R_0^+ \oplus R_1$$

where the spectrum of $R_0^+$ is *discrete* and the spectrum of $R_1$ is *continuous*. In fact the theory of Eisenstein series constructs $R_1$ *explicitly* as a direct integral of certain "principal series" representations $R(x,z)$. (Cf. Theorem 8.8.) This same theory however tells us little about which representations occur *in* $R_0^+$ and it is at *this* problem that the Selberg trace formula is directed.

Suppose we regard $R$ as a representation of the group algebra $L^1(Z_\infty^+\backslash G_{\mathbf{A}})$ by defining $R(f)$, as usual, by

$$R(f)h(y) = \int_{Z_\infty^+\backslash G_{\mathbf{A}}} f(x)(R(x)h)(y)\,dx \ .$$

Then for all $h \in L^2(Z_\infty^+ G_{\mathbf{Q}}\backslash G_{\mathbf{A}})$,

$$(9.18) \qquad (R(f)h)(y) = \int_X K(x,y)h(x)\,dx$$

where

$$(9.19) \qquad K(x,y) = \sum_{\gamma \in G_{\mathbf{Q}}} f(x^{-1}\gamma y) \ .$$

In Subsection B below we shall prove that $R_0^+(f)$ (not $R(f)$!) is of trace class for suitably regular $f$. The trace formula will then amount to finding a useful expression for this trace by realizing $R_0^+(f)$ as an integral operator with kernel $K_0^+(x,y)$ whose trace is simply the integral of this kernel "over the diagonal."

Note that $K_0^+ = K - K_1$ where $K_1$ is the kernel of $R_1(f)$, the restriction of $R(f)$ to the subspace $L_1^2(E)$. Therefore

$$\operatorname{tr}(R_0^+(f)) = \int_X K_0^+(x,x)\,dx$$

$$= \int_X \{K(x,x) - K_1(x,x)\}\,dx$$

and the theory of Eisenstein series enters because the kernel $K_1$ is described explicitly in terms of Eisenstein series! (Cf. Theorem 8.8 again.)

To obtain a more useful form of the trace formula (still however a "first" form of the formula!) one has to break up $K(x,x)$ and $K_1(x,x)$ into a number of terms and then evaluate the integrals which result from regrouping these terms in some sensible fashion. This will be done in Subsection C.

It must be emphasized that although the form of the trace formula ultimately given in Theorem 9.22 is sufficient for many purposes (in particular for the applications described in Section 10) it falls far short of an "explicit" trace formula analogous to either the Poisson summation formula (for the real line) or formula (9.16) (for the case of compact quotient).

### B. The Trace of $R_0^+(f)$

The burden of this subsection is to prove that $R_0^+(f)$ is an integral operator of trace class *for suitably regular* f and that its trace is given by the integral of its kernel over the diagonal. This is the content of Theorem 9.7 which constitutes a preliminary form of (the first form of) the Selberg trace formula.

THROUGHOUT THE REMAINDER OF SECTION 9 THE FOLLOWING ASSUMPTION WILL BE IN FORCE:

ASSUMPTION 9.1. The function f is the convolution of two bi-K-finite functions $f'$ and $f''$ in $C_c^\infty(Z_\infty^+\backslash G_A)$.

Recall that

$$R_0^+(f) = R(f) - R_1(f)$$

where $R_1(f) = R(f)E_1E_1^*$ (Theorem 8.8). To exhibit $R_0^+(f)$ as an integral operator in $L^2(X)$ it therefore suffices to separately analyze $R(f)$ and $R_1(f)$. But the operator $R(f)$ is given *a priori* as an integral operator with kernel

$$K(x,y) = \sum_{\gamma \in G_Q} f(x^{-1}\gamma y) \ . \tag{9.20}$$

Furthermore this kernel is a smooth function on $X \times X$ since for $x, y$ lying in *compact* subsets of $X$ the sum in (9.20) is actually finite. Thus it remains only to analyze $R_1(f)$.

The kernel of $R_1(f)$ being non-trivial to describe we shall first introduce certain R-invariant subspaces $L_1^2(T)$ of $L_1^2(E)$. Then we shall obtain a formula for the kernel of $R_1(f)$ restricted to each $L_1^2(T)$ and "sum up" the resulting integral operators to describe $R_1(f)$ itself.

Fix $T$ an arbitrary positive number and let $\hat{L}_1^2(T)$ denote the subspace of $\hat{L}_1^2(E) = \bigoplus_{(\tau)} \hat{L}_1^2(E,\tau)$ consisting of functions whose projection onto any of the summands $L_1^2(E,\tau)$ has (compact) support in $[-iT, iT]$. Then the image

$$L_1^2(T) = E_1(\hat{L}_1^2(T))$$

is a R-invariant subspace of $L^2(X)$ and the integral operator $R_1(f, T)$ corresponding to it is defined by the composition of $R(f)$ with the projection $E_{1,T}E_{1,T}^*$ of $L^2(X)$ onto $L_1^2(T)$.

To describe the kernel of $R_1(f, T)$ we need to fix an orthonormal basis $\{\phi_j\}_{j\epsilon I}$ for $H$ which is compatible with the decomposition $H = \oplus(\tau)$. We denote by $R_{ij}(f,z)$ the matrix coefficient $(R(f,z)\phi_j, \phi_i)$ of $R(f,z)$ and define $K_1(x,y,f,z)$ on $X \times X$ by

$$K_1(x,y,f,z) = \frac{1}{4\pi} \sum_{i,j \in I} R_{ij}(f,z)E(x,\phi_i,z)\overline{E(y,\phi_j,z)} \tag{9.21}$$

Note that since $f$ is bi-K-finite, $R_{ij}(f,z)$ vanishes for all but finitely many $i, j$. Thus the sum defining $K_1(x,y,f,z)$ is actually finite.

LEMMA 9.2. $R_1(f,T)$ *is an integral operator on* $L^2(X)$ *with (continuous) kernel*

$$K_1^T(x,y,f) = \int_{-iT}^{iT} K_1(x,y,f,z)\,d|z| \ .$$

*That is,*

$$R_1(f,T)h(x) = \int_X K_1^T(x,y,f)h(y)\,dy$$

*for all* $h \in C_c^\infty(X)$.

*Proof.* This lemma follows from Lemmas 8.6, 8.11 and the intertwining relation (8.44). Indeed, those identities imply

$$\begin{aligned}
(R_1(f,T)h)(x) &= R(f)E_{1,T}E_{1,T}^*h(x) \\
&= R(f)\left\{\frac{1}{2\pi}\int_{-iT}^{iT} E(x, E_{1,T}^*h(z), z)\,d|z|\right\} \\
&= \frac{1}{2\pi}\int_{-iT}^{iT} E(x, R(f,z)E_{1,T}^*h(z), z)\,d|z| \\
&= \frac{1}{2\pi}\int_{-iT}^{iT} E\Big(x, \sum_{j\in I}(E_{1,T}^*h(z),\phi_j)R(f,z)\phi_j, z\Big)\,d|z| \\
&= \frac{1}{2\pi}\int_{-iT}^{iT} E\Big(x, \sum_j (E_{1,T}^*h(z),\phi_j)\Big)\sum_i (R(f,z)\phi_j,\phi_i)\phi_i, z)\,d|z| \\
&= \frac{1}{2\pi}\int_{-iT}^{iT} \sum_{i,j\in I}(E_{1,T}^*h(z),\phi_j)R_{ij}(f,z)E(x,\phi_i,z)\,d|z| \\
&= \int_{-iT}^{iT} \frac{1}{4\pi}\sum_{i,j}\Big(\int_X h(y)\overline{E(y,\phi_j,z)}\,dy\Big) R_{ij}(f,z)E(x,\phi_i,z)\,d|z| \\
&= \int_X\Big(\int_{-iT}^{iT}\frac{1}{4\pi}\sum_{i,j}R_{ij}(f,z)E(x,\phi_i,z)\overline{E(y,\phi_j,z)}\Big)h(y)\,dy \ ,
\end{aligned}$$

as desired. □

The problem of describing the kernel of $R_1(f)$ is now surmounted by the following lemma.

LEMMA 9.3. *Given any Siegel domain* $\gamma(c)$, *there exist constants* C *and* M *such that*

$$(9.22)\qquad \int_{-i\infty}^{i\infty} |K_1(x,y,f,z)|\, d|z| \le Ce^{-MH(x)}e^{-MH(y)}$$

*for all* $x, y \in \gamma(c)$.

*Proof.* From Schwartz' Lemma (and some straightforward manipulation) it follows that, for $z \in i\mathbb{R}$,

$$(9.23)\qquad |K_1(x,y,f,z)| \le K_1(x,x,{}^1f,z)^{1/2}\, K_1(y,y,{}^2f,z)^{1/2} .$$

Here ${}^1f = f' * (f')^*$ and ${}^2f = (f'')^* * f''$ if $f = f' * f''$. (Recall that $f^*(y) = f(y^{-1})$.) Therefore, to prove this lemma, it suffices to establish (9.22) for $f = {}^1f$ (or ${}^2f$) and $y = x$. (Using (9.23) simply apply Cauchy-Schwartz to the integral in (9.22)!)

But by Lemma 9.2,

$$\int_{-iT}^{iT} K_1(x,x,{}^1f,z)\, d|z|$$

is the value on the diagonal of the kernel of $R_1({}^1f, T)$. Since $R_1({}^1f,T)$ is the restriction to $L_1^2(T)$ of the positive semi-definite operator $R({}^1f)$, its kernel can be bounded on the diagonal by the kernel of $R({}^1f)$, i.e.

$$\int_{-iT}^{iT} K_1(x,x,{}^1f,z)\, d|z| \le K(x,x) ,$$

where

$$K(x,x) = \sum_{\gamma \in G_Q} {}^1f(x^{-1}\gamma x) .$$

But $|K(x,x)|$, in turn, may be bounded in $\gamma(c)$ by some $Ce^{-MH(x)}$, and this bound is independent of T. ([Harish-Chandra 2], p. 9.) Therefore, since $|K_1(x,x,{}^1f,z)| = K_1(x,x,{}^1f,z)$, the lemma follows. □

PROPOSITION 9.4. $R_1(f)$ *is an integral operator with kernel*

$$K_1(x,y) = \int_{-i\infty}^{i\infty} K_1(x,y,f,z)d|z| \tag{9.24}$$

*Proof.* Lemmas 9.2 and 9.3. □

COROLLARY 9.5. $R_0^+(f)$ *is an integral operator in* $L_0^2(X)$ *with kernel* $K_0^+(x,y) = K(x,y) - K_1(x,y)$.

PROPOSITION 9.6. *The operator* $R_0^+(f)$ *possesses a trace.* (*Remember we are assuming* f *satisfies* Assumption 9.1.)

*Proof.* For any $h \in C_c^\infty(Z_\infty^+\backslash G_A)$ it is known that $R_0(h)$ is of Hilbert-Schmidt type, i.e. $\sum_i \|R_0(h)\phi_i\|^2 < \infty$ for any complete orthonormal system $\phi_i$ in $L_0^2(X)$. This "refinement" of Lemma 5.2 is an immediate consequence of the estimate (5.4). Indeed (5.4) implies that $R_0(h)\phi(x) = (k_x,\phi)$, with $k_x \in L_0^2(X)$ and $\|k_x\|_2 < c$ (some constant independent of $\phi$). So if $k(x,y) = k_x(y)$,

$$R_0(h)\phi(x) = \int_X k(x,y)\phi(y)dy$$

with $k(x,y) \in L^2(X\times X)$. Thus $R_0(h)$ is Hilbert-Schmidt by a standard theorem of functional analysis.

Now recall that $f = f' * f''$ with both $f'$ and $f''$ in $C_c^\infty(Z_\infty^+\backslash G_A)$. Therefore $R_0(f)$ is of trace class (being the product of two Hilbert-Schmidt operators) and it remains only to see that $R^+(f)$ is also. But this is immediate from Theorem 8.8. Indeed the K-finiteness of f implies that all but finitely many of the summands

$$\int_X f(x) D_\chi(x) I \, dx$$

of $R^+(f)$ coincide with the zero operator. Thus $R^+(f)$ is of finite rank let alone trace class! □

THEOREM 9.7. *The kernel* $K_0^+(x,y)$ *is integrable "over the diagonal" and its integral equals the trace of* $R_0^+(f)$. *That is*

$$\text{tr}\,(R_0^+(f)) = \int_X K_0^+(x,x)\,dx \ . \tag{9.25}$$

*Proof.* If we had already shown $K_0^+(x,y)$ to be continuous the conclusion above would be an immediate consequence of the well-known lemma which asserts that if a trace class operator is represented by a *continuous* square-integrable kernel on $X \times X$ then the kernel is integrable on $X$ and its integral gives the trace of this operator. Since we have *not* already shown $K_0^+(x,y)$ to be continuous we shall proceed more deviously as follows.

Observe first that $K_0^+(x,y)$ is at least continuous in x and y separately. Indeed since $K(x,y)$ is continuous, it suffices to deal with $K_1(x,y)$. From the proof of Lemma 9.3, recall that

$$|K_1(x,y,f,z)| \le K_1(x,x,{}^1f,z)^{1/2} K(y,y,{}^2f,z)^{1/2} \ ,$$

so from Lemma 9.3 itself it follows that (for y in any Siegel domain)

$$\int_{\substack{z \in i\mathbf{R} \\ |z|>T}} |K_1(x,y,f,z)|\,d|z| \le Ce^{-MH(y)} \left( \int |K_1(x,x,{}^1f,z)|\,d|z| \right)^{1/2} .$$

This means that for any fixed $x$, the integral defining $K_1(x,y)$ converges uniformly for $y$ in compact subsets of $X$. Therefore this integral defines a continuous function of $y$ (and a similar argument works for $x$).

Now let $H'(\,,\,)$ and $H''(\,,\,)$ denote the kernels representing the Hilbert-Schmidt operators $R_0^+(f')$ and $R_0^+(f'')$. According to a well-known principle from functional analysis, the function

$$H(x,y) = \int_X H'(x,v)H''(v,y)dv$$

is defined almost everywhere and determines the Hilbert-Schmidt kernel of $R_0^+(f)$. Moreover,

$$\operatorname{tr} R_0^+(f) = \int_X H(x,x)dx \;.$$

Therefore to prove the theorem it will suffice to construct $H'$ and $H''$ such that $H(x,x) = K_0(x,x)$ for almost all $x$.

Actually, since $K_0(x,y)$ must *a priori* equal $H(x,y)$ almost everywhere on $X \times X$, and since $K_0(x,y)$ has just been seen to be separately continuous in $x$ and $y$, it will suffice to find $H'$ and $H''$ so that $H(x,y)$ too is separately continuous.

So recall again that for any K-finite $f \in C_c^\infty(Z_\infty^+\backslash G_{\mathbf{A}})$ and Siegel domain $\mathcal{Y}(c)$, there exist $C$ and $M$ so that

$$\left|\sum_{\gamma \in G_{\mathbf{Q}}} f(x^{-1}\gamma v)\right| \le Ce^{-MH(x)}, \quad x, v \in \mathcal{Y}(c) \;. \tag{9.26}$$

For any fixed $x$ it follows that the linear form on $H_0^+ = L_0^2(X) \oplus (\oplus D_\chi)$ defined by

$$\phi \mapsto \int_X \sum_{\gamma \in G_{\mathbf{Q}}} f(x^{-1}\gamma v)\phi(v)dv$$

is continuous. Hence there is a unique function $h_x'(v)$ in $H_0^+$ such that

$$\int_X \sum_{\gamma \epsilon G_Q} f(x^{-1}\gamma v)\phi(v)\,dv = \int_X h'_x(v)\phi(x)\,dv \;,$$

for all $\phi \epsilon H_0^+$, i.e.

$$H'(x,v) = h'_x(v)$$

is a Hilbert-Schmidt kernel for $R_0^+(f')$ on $L^2(X)$.

In a similar way the Hilbert-Schmidt kernel $H''(v,y)$ of $R_0^+(f'')$ may be introduced so that $H''(v,y)$ also belongs to $H_0^+$ for each y. Setting

$$H(x,y) = \int_X H'(x,v)H''(v,y)\,dv$$

then describes the desired Hilbert-Schmidt kernel of $R_0^+(f)$. Indeed from (9.26) and the dominated convergence theorem comes the required conclusion that $H(x,y)$ is continuous in x for any fixed y (and similarly in y for any fixed x).

This completes the proof of Theorem 9.7.

## C. A Second Form of the Trace Formula

Recall that for f satisfying Assumption 9.1 the trace of $R_0^+(f)$ exists and equals

$$\int_X K_0^+(x,x)\,dx$$

where $K_0^+(x,x) = K(x,x) - K_1(x,x)$,

$$K(x,y) = \sum_{\gamma \epsilon G_Q} f(x^{-1}\gamma y) \tag{9.27}$$

and $K_1(x,y)$ is described by (9.21) and (9.24).

A "second form" of the trace formula (Theorem 9.22) will now be obtained by

(i) breaking $K(x,x)$ up into elliptic and parabolic parts;

(ii) writing $K_1(x,x)$ as the sum of a part "at infinity" plus its complement;

(iii) regrouping these terms in a meaningful way;

and

(iv) integrating the resulting components.

The work involved will be tedious but the plan of attack is quite simple.

1. *Conjugacy Classes in* $G_Q$

We shall say that $\gamma \in G_Q$ is *parabolic* if it is $G_Q$-conjugate to some element of $B_Q$ and *elliptic* otherwise. The set of all elliptic elements in $G_Q$ will be denoted by $G_e$. The parabolic conjugacy classes may be classified as follows.

If $\mu = \begin{bmatrix} \mu_1 & 0 \\ 0 & \mu_2 \end{bmatrix}$ belongs to $A_Q$ we say that $\mu$ is *regular* if $\mu_1 \neq \mu_2$ and *singular* otherwise. The singular elements of $A_Q$ then coincide with $Z_Q$. The set of regular elements in $A_Q$ will be denoted by $A_r$. Note that

$$\begin{bmatrix} \mu_1 & n \\ 0 & \mu_2 \end{bmatrix} = \mu \begin{bmatrix} 1 & n\mu_1^{-1} \\ 0 & 1 \end{bmatrix} = \begin{bmatrix} 1 & \frac{n}{\mu_1-\mu_2} \\ 0 & 1 \end{bmatrix}^{-1} \mu \begin{bmatrix} 1 & \frac{n}{\mu_1-\mu_2} \\ 0 & 1 \end{bmatrix}$$

if $\mu = \begin{bmatrix} \mu_1 & 0 \\ 0 & \mu_2 \end{bmatrix} \in A_r$. It follows that every *parabolic* $\gamma$ in $G_Q$ is conjugate either to some

(1) $\mu \in Z_Q$;

(2) $\mu \in A_r$; or

(3) $\mu\nu$, with $\mu \in Z_Q$ and $\nu \neq e$ in $N_Q$. Thus $A_r$ and $Z_Q N_Q$ describe the parabolic conjugacy classes in $G_Q$.

We shall now decompose $K(x,x)$ by replacing the sum over $G_Q$ in (9.27) by a sum over the various orbits of $G_Q$ defined by conjugation.

If $H$ is any $Q$-subgroup of $GL(2)$ let $H(\gamma)$ denote the centralizer of $\gamma$ in $H$. If $\{G_e\}$ denotes a fixed set of representatives of $G_Q$-conjugacy $G_Q$-conjugacy classes in $G_Q$ then $K(x,x)$ may be expressed as a sum of the following terms:

(i) *the elliptic contribution*

(9.28) $$I_e(x,f) = \sum_{\gamma\in\{G_e\}} \sum_{\delta\in G(\gamma)_Q\backslash G_Q} f(x^{-1}\delta^{-1}\gamma\delta x);$$

(ii) *the singular contribution*

(9.29) $$I_s(x,f) = \sum_{\mu\in Z_Q} f(\mu);$$

(iii) *the contribution from* $A_r$:

(9.30) $$\sum_{\delta\in N(A)_Q\backslash G_A} \sum_{\mu\in A_r} f(x^{-1}\delta^{-1}\mu\delta x)$$

where $N(A)$ denotes the normalizer of $A$; and

(iv) *the contribution from* $Z_Q\backslash N_Q$:

(9.31) $$\sum_{\delta\in B_Q\backslash G_Q} \sum_{\mu\in Z_Q} \sum_{\substack{\nu\in N_Q\\ \nu\neq e}} f(x^{-1}\delta^{-1}\mu\nu\delta x).$$

The last three terms above comprise the *parabolic part of* $K(x,x)$. In describing the contribution from $A_r$ alone we have used the fact that $\delta^{-1}\mu\delta = \delta_1^{-1}\mu_1\delta_1$ (with $\mu,\mu_1$ in $A_r$) implies $\delta = y\delta_1$ for some $y\in N(A)_Q$. Similarly for (9.31) we used the fact that $\delta^{-1}\mu\nu\delta = \delta_1^{-1}\mu_1\nu_1\delta_1$ implies $\delta = b\delta_1$ for some $b\in B_Q$.

2. *Truncating* $K_1(x,x)$ *and* $K(x,x)$

To analyze $K_1(x,x)$ and to *further* decompose and rearrange $K(x,x)$ the following lemma will be crucial.

LEMMA 9.8. *There exists a constant* $c_1 > 1$ *with the property that* $\gamma\in B_Q$ *if and only if* $H(\gamma x) > \log c_1$ *for all* $x\in G_A$ *such that* $H(x) > \log c_1$.

*Proof.* If "only if" direction is trivial. Indeed $H\left(\begin{bmatrix}\alpha & 0\\ 0 & \beta\end{bmatrix}\right) = 1$ for all rational $\alpha, \beta$ and therefore any $c_1 > 1$ will do.

Now choose $c_1 > 1$ so that $H(wn) < \log c_1$ for all $n \in N_{\mathbf{A}}$ (recall that $w = \begin{bmatrix} 0 & 1\\ -1 & 0\end{bmatrix}$). Then for the "if" part suppose $\gamma \notin B_{\mathbf{Q}}$. By Bruhat's decomposition

$$\gamma = bwn_\gamma \quad \text{with} \quad b \in B_{\mathbf{Q}} \quad \text{and} \quad n_\gamma \in N_{\mathbf{Q}} .$$

So if $x = na\, h_t\, kz$ (cf. (8.8)),

$$wn_\gamma x = wn_\gamma nah_t\, kz = (h_t^{-1}a^{-1})wn_1\, kz$$

where

$$n_1 = h_t^{-1}a^{-1}n_\gamma nah_t$$

belongs to $N_{\mathbf{A}}$. Therefore

$$H(\gamma x) = H(wn_\gamma x) = H(h_t^{-1})H(wn_1) ,$$

which implies $H(\gamma x) < \log c_1$, a contradiction. □

To explain the significance of this lemma it is necessary to introduce the set

$$S(c) = \{x \in G_{\mathbf{A}} : H(x) \geq \log c\} .$$

This set is a kind of extended Siegel domain. Its projection onto $X$ is the set of points in the fundamental domain which are sufficiently close to infinity (in the upper half plane these points comprise a vertical strip truncated from below at some fixed positive height). From reduction theory (cf. [Godement 3]) it is known that

(i) for some $c > 0$, $G_{\mathbf{A}} = G_{\mathbf{Q}}S(c)$; and

(ii) for any $c > 0$, the set of $\gamma \in G_{\mathbf{Q}}$ such that $S(c) \cap \gamma S(c) \neq \phi$ is finite modulo $B_{\mathbf{Q}}$.

Lemma 9.8 is important precisely because *for* $\varepsilon > c_1$ the set $S(\varepsilon)$ is particularly well-suited to the further analysis of $K_1(x, y)$.

Indeed suppose $\varepsilon > c_1$. If we let $\chi_\varepsilon(x)$ denote the characteristic function of $S(\varepsilon)$ then by (ii) above the series

$$\sum_{\delta \in B_Q \backslash G_Q} E^0(\delta x, \phi, z)\chi_\varepsilon(\delta x) \tag{9.32}$$

*is a finite sum for* $x \in S(c)$. In particular (9.32) defines a function on $Z_\infty^+ B_Q \backslash G_A$ which we shall denote by $E'_\varepsilon(x, \phi, z)$. The point is that Lemma 9.8 above implies that

$$E'_\varepsilon(x, \phi, z) = E^0(x, \phi, z)$$

*for all* $x \in S(\varepsilon)$. Here, as always, $E^0$ denotes the "constant term" of E.)

Using $E'_\varepsilon(x, \phi, z)$ we introduce the "adjusted" kernel

$$K'(x, f, \varepsilon) = \frac{1}{4\pi}\int_{-i\infty}^{i\infty} \sum_{i,j} R_{ij}(z, f)\, E'_\varepsilon(x, \phi_i, z)\overline{E'_\varepsilon(x, \phi_j, z)}\, d|z| . \tag{9.33}$$

This is the "part at infinity" of $K_1(x, x)$. Its "complement" is

$$K''(x, f, \varepsilon) = K_1(x, x) - K'(x, f, \varepsilon) . \tag{9.34}$$

Note that the right hand side of (9.33) *does* converge. In fact from Lemma 9.8 it follows that

$$K'(x,f,\varepsilon) = \sum_{\delta \in B_Q \backslash G_Q} \int_{-i\infty}^{i\infty} \frac{1}{4\pi}\sum_{i,j} R_{ij}(z,f)E^0(x\delta,\phi_i,z)\overline{E^0(x\delta,\phi_j,z)}d|z|\chi_\varepsilon(x\delta). \tag{9.35}$$

Therefore the absolute value of $K'(x, f, \varepsilon)$ is bounded above (in any $S(c)$) by $Ce^{-MH(x)}$ by Lemma 9.3 (recall that $E^0(x, \phi, z)$ is obtained by integrating $E(xn, \phi, z)$ over the *compact* set $N_Q \backslash N_A$).

REMARK. We introduce these functions because $K(x, x)$ is *not* integrable over X yet $K''(x, f, \varepsilon)$ is. Thus $K'(x, f, \varepsilon)$ isolates the "bad part" of $K(x, x)$.

Now let us re-examine the (non-integrable) contributions of $A_r$ and $A_s N_Q$ to the parabolic part of $K(x,x)$. Since $A_Q$ is of index two in $N(A)_Q$ the contribution from $A_r$ becomes

$$\frac{1}{2} \sum_{\delta \in A_Q \backslash G_Q} \sum_{\mu \in A_r} f(x^{-1}\delta^{-1}\mu\delta x)$$

which we decompose as the sum of

$$J_r(x,f,\varepsilon) = \frac{1}{2} \sum_{\mu \in A_r} \sum_{\delta \in A_Q \backslash G_Q} f(x^{-1}\delta^{-1}\mu\delta x)(\chi_\varepsilon(\delta x) + \chi_\varepsilon(w\delta x)) \tag{9.36}$$

and

$$I_r(x,f,\varepsilon) = \frac{1}{2} \sum\sum f(x^{-1}\delta^{-1}\mu\delta x)(1-\chi_\varepsilon(\delta x) - \chi_\varepsilon(w\delta x)) \ . \tag{9.37}$$

Similarly we express the contribution from $Z_Q N_Q$ as the sum of

$$J_s(x,f,\varepsilon) = \sum_{\mu \in Z_Q} \sum_{\delta \in B_Q \backslash G_Q} \sum_{\substack{\nu \in N_Q \\ \nu \neq e}} f(x^{-1}\delta^{-1}\mu\nu\delta x)\chi_\varepsilon(\delta x) \tag{9.38}$$

and

$$I_s(x,f,\varepsilon) = \sum\sum\sum f(x^{-1}\delta^{-1}\mu\nu\delta x)(1-\chi_\varepsilon(\delta x)) \ . \tag{9.39}$$

Recalling that $K_0^+(x,x) = K(x,x) - K_1(x,x)$ we finally rewrite $K_0^+(x,x)$ as the sum of

(9.40)

$I_e(x,f)$ (the elliptic term) ,

$I_s(x,f)$ (the singular term),

$J_r(x,f,\varepsilon) + J_s(x,f,\varepsilon) - K'(x,f,\varepsilon)$ (the first parabolic term) ,

$I_r(x,f,\varepsilon) + I_s(x,f,\varepsilon)$ (the second parabolic term) ,

and

$-K''(x,f,\varepsilon)$ (the third parabolic term) .

This then is the desired decomposition and rearrangement of the kernel $K_0^+(x,x)$. The $\varepsilon$-neighborhood of infinity $S(\varepsilon)$ was used to separate the integrable and non-integrable parts of $K(x,x)$ and $K_1(x,x)$.

### 3. *Plan of Attack*

Our final form for the trace formula will arise from evaluating the integral (over $X$) of each of the five terms of (9.39). Precisely because of how we have defined and grouped the components $K'$, $K''$, $J_r$, $I_r$, $J_s$, and $I_s$, it will turn out that *each* of the five terms of (9.39) (and not just their sum) *is* integrable. Indeed the non-integrable parts of $K(x,x)$ and $K_1(x,x)$ have been isolated in the first parabolic term and essentially cancel each other out there. Our plan of attack is simply to deal with each of these five terms individually.

4. *The Elliptic and Singular Terms*

The elliptic term $I_e(x,f)$ is given by the series

$$\sum_{\gamma \in G_e} f(x^{-1}\gamma x) .$$

To prove that $I_e(x,f)$ is integrable it will therefore suffice to prove that the function

$$F(x) = \sum_{\gamma \in G_e} |f(x^{-1}\gamma x)| \quad (z \in Z_\infty^+ G_Q \backslash G_A)$$

is compactly supported.

LEMMA 9.9. *Suppose* $C$ *is a compact subset of* $G_A$ *modulo* $Z_\infty^+$. *Then there exists a number* $d_C$ *with the following property: If* $\gamma$ *in* $G_Q$ *is such that*

$$x^{-1}\gamma x \in C$$

*for some* $x \in G_A$ *with* $H(x) > \log d_C$ *then* $\gamma \in B_Q$.

*Proof.* If $x = nak$, then the hypothesis $x^{-1}\gamma x \in C$ implies that

$$a^{-1}n^{-1}\gamma\, na \in C' \tag{9.41}$$

where $C' = KCK$. But

$$\begin{bmatrix} \mu_1 & * \\ 0 & \mu_2 \end{bmatrix}^{-1} \begin{bmatrix} a & b \\ c & d \end{bmatrix} \begin{bmatrix} \mu_1 & * \\ 0 & \mu_2 \end{bmatrix} = \begin{bmatrix} * & * \\ \frac{\mu_1 c}{\mu_2} & * \end{bmatrix}$$

if $\gamma = \begin{bmatrix} a & b \\ c & d \end{bmatrix}$. Consequently (9.41) implies that $|c\mu_1\mu_2^{-1}|$ must be bounded, i.e.

$$|c| < M\left|\frac{\mu_1}{\mu_2}\right|^{-1} .$$

Therefore if $\left|\frac{\mu_1}{\mu_2}\right|^{-1}$ is sufficiently small, i.e. if $H(x) = \log\left|\frac{\mu_1}{\mu_2}\right|^{1/2}$ is sufficiently *large*, c must be zero (which means $\gamma = \begin{bmatrix} a & b \\ c & d \end{bmatrix}$ belongs to $B_Q$). □

From this lemma it follows immediately that $F(x) = \sum_{\gamma \in G_e} |f(x^{-1}\gamma x)|$ is compactly supported and hence that $I_e(x,f)$ is (absolutely) integrable. Furthermore its integral over X equals

$$\int_{Z_\infty^+ G_Q \backslash G_A} \sum_{\gamma \in \{G_e\}} \sum_{\delta \in G(\gamma)_Q \backslash G_Q} f(x^{-1}\delta^{-1}\gamma\delta x)\,dx$$

$$= \sum_{\gamma \in \{G_e\}} \int_{Z_\infty^+ G(\gamma)_Q \backslash G_A} f(x^{-1}\gamma x)\,dx .$$

Therefore

$$\int I_e(x,f)\,dx = \sum_{\gamma \in \{G_e\}} \mathrm{meas}(Z_\infty^+ G(\gamma)_Q \backslash G(\gamma)_A) \int_{G(\gamma)_A \backslash G_A} f(x^{-1}\gamma x)\,dx .$$

As for the singular term, there is little to say except that X has finite measure. Summing up, we have:

PROPOSITION 9.10 (Contribution of the elliptic and singular terms to the Trace Formula). *The terms* $I_e(x,f)$ *and* $I_s(x,f)$ *are integrable over* X *and their integral equals*

$$\mathrm{meas}(Z_\infty^+ G_Q \backslash G_A) \sum_{\mu \in Z_Q} f(\mu) + \sum_{\gamma \in \{G_e\}} \mathrm{meas}(Z_\infty^+ G(\gamma)_Q \backslash G(\gamma)_A) \int_{G(\gamma)_A \backslash G_A} f(x^{-1}\gamma x)\,dx.$$

5. *The First Parabolic Term*

This term equals

$$J_r(x,f,\varepsilon) + J_s(x,f,\varepsilon) - K'(x,f,\varepsilon) \ . \tag{9.42}$$

Although each expression in (9.42) is *non-integrable* over $X$ (in fact isolates the "bad part" of either $K(x,x)$ or $K_1(x,x)$) the *sum* of these three expressions *is* integrable and its integral approaches $0$ as $\varepsilon \to 0$. To see this we need to fiddle a bit with $J_r(x,f,\varepsilon)$ and $J_s(x,f,\varepsilon)$.

Recall first that

$$J_r(x,f,\varepsilon) = \frac{1}{2} \sum_{\mu \in A_r} \sum_{\delta \in A_Q \backslash G_Q} f(x^{-1}\delta^{-1}\mu\delta x)(\chi_\varepsilon(\delta x) + \chi\varepsilon(w\delta x)) \ .$$

Since $w$ normalizes $A_r$, it is easy to show that

$$J_r(x,f,\varepsilon) = \sum_{\mu \in A_r} \sum_{\delta \in A_Q \backslash G_Q} f(x^{-1}\delta^{-1}\mu\delta x)\chi_e(x\delta)$$

$$= \sum_{\mu \in A_r} \sum_{\delta \in N_Q A_Q \backslash G_Q} \sum_{\nu \in N_Q} f(x^{-1}\delta^{-1}\mu\nu\delta x)\chi_\varepsilon(x\delta)$$

(since $\begin{bmatrix} 1 & n \\ 0 & 1 \end{bmatrix}^{-1} \begin{bmatrix} a_1 & 0 \\ 0 & a_2 \end{bmatrix} \begin{bmatrix} 1 & n \\ 0 & 1 \end{bmatrix} = \begin{bmatrix} a_1 & 0 \\ 0 & a_2 \end{bmatrix} \begin{bmatrix} 1 & \frac{n(a_1-a_2)}{a_1} \\ 0 & 1 \end{bmatrix}$). On the other hand $J_s(x,f,e)$ is the difference of

$$\sum_{\mu \in Z_Q} \sum_{\delta \in B_Q \backslash G_Q} \sum_{\nu \in N_Q} f(x^{-1}\delta^{-1}\mu\nu\delta x)\chi_\varepsilon(x\delta)$$

and

$$\sum_{\mu \in Z_Q} \sum_{\delta \in B_Q \backslash G_Q} f(x^{-1}\delta^{-1}\mu\delta x)\chi_\varepsilon(x\delta) \ . \tag{9.43}$$

LEMMA 9.11. *The integral* (over $X$) *of the absolute value of (9.43) approaches* $0$ *as* $\varepsilon \to \infty$.

*Proof.* The integral of the absolute value of (9.42) is bounded by

$$\sum_{\mu \in Z_Q} \int_{Z_\infty^+ B_Q \backslash G_A} |f(x^{-1}\mu x)|\, \chi_\varepsilon(x)\, dx$$

which (by (8.11)) equals

$$\sum_{\mu \in Z_Q} c_G \left( \int_{\log \varepsilon} e^{-2t} dt \right) \left( \int_{N_Q \backslash N_A} \int_K |f(k^{-1} n^{-1} \mu n k)|\, dn\, da\, dk \right) .$$

Since the support of $f$ is compact modulo $Z_\infty^+$, the double integral above is non zero for only finitely many $\mu \in Z_Q$ ($N_Q \backslash N_A$ is compact!). Thus the lemma follows immediately. □

REMARK 9.12. In the sequel we shall encounter several terms which (like (9.43)) depend on $\varepsilon$ and approach $0$ (in $L^1$-norm) as $\varepsilon \to \infty$. Since $\varepsilon$ was chosen *arbitrarily* larger than the $c_1$ of Lemma 9.8, such terms contribute nothing to the final trace formula and therefore can be (and will be) discarded.

From Lemma 9.11 it follows that

$$(9.44) \quad J_r(x,f,\varepsilon) + J_s(x,f,\varepsilon) = \sum_{\mu \in A_Q} \sum_{\delta \in B_Q \backslash G_Q} \left( \sum_{\nu \in N_Q} f(x^{-1}\delta^{-1}\mu\nu\delta x) \right) \chi_\varepsilon(x\delta)$$

*modulo* an expression whose integral over $X$ approaches $0$ as $\varepsilon \to \infty$. Before putting this expression together with $-K'(x,f,\varepsilon)$ let us further refine (9.44).

Note that (for each $y \in G_A$ and $\mu \in A_Q$) the function

$$u \to f\left(y^{-1}\mu \begin{bmatrix} 1 & u \\ 0 & 1 \end{bmatrix} y\right)$$

is of Schwartz-Bruhat type on $\mathbf{A}$. Thus we might hope to apply Poisson summation to the inner sum in (9.44). To this end, fix once and for all a non-trivial unitary character $\tau$ of $Q\backslash A$. For each $\xi \in A$ define

$$\hat{f}(\xi) = \hat{f}(\xi,\mu,y) = \int_A f\left(y^{-1}\mu\begin{bmatrix}1 & u\\0 & 1\end{bmatrix}y\right)\tau(\xi u)\,du\ .$$

LEMMA 9.13. *Modulo a function whose integral over* X *approaches* 0 *as* $\varepsilon \to \infty$,

$$J_r(x,f,\varepsilon) + J_s(x,f,\varepsilon) = \sum_{\delta \in B_Q\backslash G_Q} \sum_{\mu \in A_Q} \hat{f}(0,\mu,\delta x)\chi_\varepsilon(\delta x)\ .$$

*Proof.* By Poisson summation on A,

$$\left\{\sum_{\nu \in N_Q} f(x^{-1}\delta^{-1}\mu\nu\delta x)\right\}\chi_\varepsilon(\delta x) = \hat{f}(0,\mu,\delta x)\chi_\varepsilon(\delta x) + \sum_{\xi \in Q^\times} \hat{f}(\xi,\mu,\delta x)\chi_\varepsilon(\delta x)\ .$$

Therefore by (9.44) it will suffice to prove that the integral (over X) of

$$\sum_{\delta \in B_Q\backslash G_Q} \sum_{\mu \in A_Q} \sum_{\xi \in Q^\times} \hat{f}(\xi,\mu,\delta x)\chi_\varepsilon(\delta x)$$

approaches 0 as $\varepsilon \to +\infty$. But this latter integral is bounded by

$$\int_{Z_\infty^+ B_Q\backslash G_A} \left(\sum_{\mu \in A_Q} \sum_{\xi \in Q^\times} |\hat{f}(\xi,\mu,x)|\right)\chi_\varepsilon(x)\,dx$$

which in turn equals

$$c_G \int_\omega \int_{\log \varepsilon}^{\infty} \int_K \sum_{\mu \in A_Q} \sum_{\xi \in Q^\times} |\hat{f}(\xi,\mu,vh_t k)|\, e^{-2t}\,dv\,dt\,dk$$

with $\omega$ a relatively compact fundamental domain for $B_Q$ in $N_A A_A^1$. So making the change of variables $v \to h_t v h_t^{-1}$ (recall that $d(ana^{-1}) = \delta_B(a)dn$) allows us to bound the integral in question by

$$c_G \int_{\omega \times K} \int_{\log \varepsilon}^{\infty} \sum_{\mu \in A_Q} \sum_{\xi \in Q^\times} |\hat{f}(\xi, \mu, h_t k)| \, dv \, dt \, dk \ . \tag{9.45}$$

(Observe that $h_t^{-1} v h_t$ is contained in $\omega$ for $t > 0$.) Now since $f$ is of compact support there are only finitely many $\mu \in A_Q$ such that $\hat{f}(\xi,\mu,vk) \neq 0$. For the same reason it follows that *for any* $N > 0$ there is a constant $C_N$ such that

$$\sum_{\xi \in Q^\times} \sum_{\mu \in A_Q} |\hat{f}(e^{2t}\xi, \mu, vk)| \leq C_N e^{-2tN}$$

for all $(v, k)$ in $\omega \times K$. (Recall that $\hat{f}(\xi, \mu, vk)$ is continuous in $vk$.)

Finally observe that $\hat{f}(\xi, \mu, h_t vk) = e^{2t} \hat{f}(e^{2t}\xi, \mu, vk)$. This implies that (9.45) is bounded by

$$C'_N \int_{\log \varepsilon}^{\infty} e^{2t} e^{-2N} \, dt \tag{9.46}$$

with $C'_N$ some constant depending on $c_G$ (and the measure of $N_Q A_Q \backslash N_A A_A^1$). So since (9.46) clearly approaches 0 as N approaches $\infty$ the lemma is proved. ❑

Recall that our goal is to show that $J_r(x, f, \varepsilon) + J_s(x, f, \varepsilon) - K'(x, f, \varepsilon)$ contributes *nothing* to the trace formula, i.e. that the integral of this expression approaches 0 as $\varepsilon \to \infty$. Thus we need (finally) to deal with the contribution from $-K'(x, f, \varepsilon)$.

By definition,

$$K'(x, f, \varepsilon) = \frac{1}{4\pi} \int_{-i\infty}^{i\infty} \sum_{i,j} R_{ij}(f, z) E'_\varepsilon(x, \phi_i, z) \overline{E'_\varepsilon(x, \phi_i, z)} \, d|z| \ .$$

But by (9.35) and (8.44) (which says that $R(f, z)$ and $E$ commute)

$$K'(x,f,\varepsilon) = \frac{1}{4\pi} \sum_{\delta \in B_Q \backslash G_Q} \int_{-i\infty}^{i\infty} \sum_j E^0(x\delta, R(f,z)\phi_j, z) \overline{E^0(x\delta, \phi_j, z)} \, d|z| \, \chi_\varepsilon(x\delta) \, .$$

So substituting for $E^0$ the expression (8.28) it follows that $K'(x,f,\varepsilon)$ is the sum of the four terms

$$\text{(i)} \quad \frac{1}{4\pi} \sum_{\delta \in B_Q \backslash G_Q} \int_{-i\infty}^{i\infty} \left\{ \sum_j R(f,z)\phi_j(\delta x)\overline{\phi_j(\delta x)} \right\} d|z| \, e^{2H(\delta x)} \chi_\varepsilon(\delta x) \, ,$$

$$\text{(ii)} \quad \frac{1}{4\pi} \sum_{\delta} \int_{-i\infty}^{i\infty} \left\{ \sum_j (M(z)R(f,z)\phi_j)(\delta x)\overline{(M(z)\phi_j)(\delta x)} \right\} d|z| e^{2H(\delta x)} \chi_\varepsilon(\delta x) \, ,$$

$$\text{(iii)} \quad \frac{1}{4\pi} \sum_{\delta} \int_{-i\infty}^{i\infty} \left\{ \sum_j (M(z)R(f,z)\phi_j)(\delta x)\overline{\phi_j(\delta x)} \, e^{-2zH(\delta x)} \right\} d|z| e^{2H(\delta x)} \chi_\varepsilon(\delta x)$$

and

$$\text{(iv)} \quad \frac{1}{4\pi} \sum_{\delta} \int_{-i\infty}^{i\infty} \left\{ \sum_j R(f,z)\phi_j(x\delta) \, \overline{M(z)\phi_j(\delta x)} \, e^{2zH(\delta x)} \right\} d|z| e^{2H(\delta x)} \chi_\varepsilon(\delta x) \, .$$

Of course to justify this last step we do have to prove that each of the integrals (i) - (iv) is finite. But since they all are rather similar it will suffice to deal with the first.

LEMMA 9.14. *For any* $y \in G_A$,

$$\int_{-i\infty}^{i\infty} \left\{ \sum_j (R(f,z)\phi_j)(y)\overline{\phi_j(y)} \right\} d|z|$$

*is finite.*

*Proof.* Recall that (by Lemma 8.1) $R(x,z)$ acts on functions in $N_A A_Q A_\infty^+ \backslash G_A$ through the formula

$$R(x,z)\phi(y) = \phi(yx) = e^{(z+1)H(yx)} e^{-(z+1)H(y)} \, .$$

Therefore

$$R(f,z)\phi(x) = \int_{Z_\infty^+\backslash G_{\mathbf{A}}} f(y)R(y,z)\phi(x)\,dy$$

$$= \int_{Z_\infty^+\backslash G_{\mathbf{A}}} f(x^{-1}y)\phi(y)e^{(z+1)H(y)}e^{-(z+1)H(x)}dy$$

$$= c_G e^{-(z+1)H(x)} \int_{N_{\mathbf{A}}} \sum_{\mu\in A_{\mathbf{Q}}} \int_{-\infty}^{\infty} \int_K f(x^{-1}n\mu h_t k)\phi(k)e^{(z-1)t}dndt\,dk$$

(by (8.11))

$$= c_G \int_K P(f,z,x,k)\phi(k)\,dk$$

where

$$P(f,z,x,y) = e^{-(z+1)H(x)} \sum_{\mu\in A_{\mathbf{Q}}} \int_{N_{\mathbf{A}}} \int_{-i\infty}^{\infty} f(x^{-1}n\mu h_t y)e^{(z-1)t}\,dndt\,.$$

Keep in mind that $f$ is compactly supported modulo $Z_\infty^+$ and $P(f,z,x,y)$ is a continuous function on the compact space $(N_A A_Q A_\infty^+\backslash G_A) \times (N_A A_Q A_\infty^+\backslash G_A)$. Thus $R(f,z)$ *is a Hilbert-Schmidt operator on* $H$ *with continuous kernel* $P(f,z,x,y)$.

On the other hand it is also clear from the formula for $P(f,z,x,y)$ that for fixed $x, y,$ and $f$, $P(f,z,x,y)$ is a Schwartz-Bruhat function of $z$ in $i\mathbf{R}$ (it is the Fourier transform of a compactly supported function!). Therefore the integral

$$\int_{-i\infty}^{i\infty} P(f,z,x,y)\,d|z|$$

is certainly finite. The lemma then follows from the well-known fact that

$$\sum R(f,z)\phi_j(x)\overline{\phi_j(y)}$$

also defines a Hilbert-Schmidt kernel for $R(f,z)$ (which coincides with $P(f,z,x,y)$ since both kernels are continuous in $x$ and $y$). □

We shall now use the identity

$$K'(x,f,\varepsilon) = (i) + (ii) + (iii) + (iv)$$

to show that $J_s + J_r - K'$ contributes nothing to the trace formula.

We deal first with (i) and (ii). From the proof of the last lemma it follows that

$$(i) = \frac{1}{4\pi} \sum_{\delta \in B_Q \backslash G_Q} \int_{-i\infty}^{-\infty} \sum_{\mu \in A_Q} \int_{N_A} \int_{-\infty}^{\infty} f(x^{-1}\delta^{-1} n\mu h_t \delta x) e^{-(z+1)t} dt dn) d|z| \chi_\varepsilon(\delta x).$$

Thus by Fourier inversion

$$(i) = \frac{1}{2} \sum_{\delta \in B_Q \backslash G_Q} \sum_{\mu \in A_Q} \left( \int_{N_A} f(x^{-1}\delta^{-1} n\mu\delta x) dn \right) \chi_\varepsilon(\delta x) .$$

Similarly (ii) is given by the same expression, since $\{M(z)\phi_j\}_j$ is again an orthonormal basis for $\mathbf{H}$ and $M(z)$ intertwines $R(f,z)$ with $R(f,-z)$. The result is that *the contributions from* (i) *and* (ii) *exactly cancel the contributions from* $J_r(x,f,\varepsilon)$ *and* $J_s(x,f,\varepsilon)$ (cf. Lemma 9.13). Therefore it remains only to prove:

LEMMA 9.15. *The integrals of (iii) and (iv) over* $X$ *approach* $0$ *as* $\varepsilon \to +\infty$.

*Proof.* Let $h_\varepsilon(x)$ denote the function defined by (iii) say. If $\gamma(c)$ is a fundamental domain for $Z_\infty^+ G_Q$ in $G_A$ then the integral of $h_\varepsilon(x)$ over $X$ is bounded by

$$c_G \int_{\log c}^{\infty} \left| \int_{N_Q \backslash N_A} \int_K h_\varepsilon(n h_t k) dn dk \right| e^{2t} dt .$$

But by the definition of the inner product in $\mathbf{H}$ this last integral equals

$$\frac{1}{4\pi}\int_{\log \varepsilon}^{\infty}\left|\int_{-i\infty}^{-\infty}\sum_{j}(M(z)R(f,z)\phi_j,\phi_j)e^{-2zt}\,d|z|\right|dt\ .$$

So since the K-finiteness of f implies that the function

$$(M(z)R(f,z)\phi_j,\phi_j) \qquad (z \in i\mathbb{R})$$

is identically zero for all but finitely many j, the contour of integration above can be changed to a line $\{z : \mathrm{Re}(z) = \delta,\ \delta > 0\}$. From this the desired conclusion follows for (iii) and a similar argument applies to (iv). □

Summing up the results of this subsection we have:

PROPOSITION 9.16 (Contribution of the first parabolic term to the trace formula). *The first parabolic term contributes nothing to the trace formula.*

*Proof.* The integral (over X) of this term depends on $\varepsilon$ and approaches 0 as $\varepsilon \to \infty$. Since the trace of $R_0^+(f)$ is independent of $\varepsilon$ the proposition is obvious. □

6. *The Second Parabolic Term*

This term equals

$$I_r(x,f,\varepsilon) + I_s(x,f,\varepsilon)$$

where $I_r$ and $I_s$ are defined by (9.37) and (9.39). From some (at first formal) manipulations it will follow that both $I_r$ and $I_s$ are integrable over X (hence the notation I). Our task will be to evaluate their integrals.

We deal first with $I_r(x,f,\varepsilon)$. Applying the usual integration techniques yields the identity

$$\int_X I_r(x,f,\varepsilon)\,dx = \frac{1}{2}\sum_{\mu\in A_r}\ \int_{Z_\infty^+ A_Q\backslash G_A} f(x^{-1}\mu x)(1-\chi_\varepsilon(x)-\chi_\varepsilon(wx))\,dx$$

$$\frac{c_G}{2}\sum_{\mu\in A_r}\ \int_{A_Q\backslash B_A^1}\int_K f(k^{-1}b^{-1}\mu bk)\int_{-\infty}^{\infty} 1-\chi_\varepsilon(bh_t)-\chi_\varepsilon(wbh_t)\,dt\ \ d_\ell b\,dk$$

since on $B_A^1$ the modular function $\delta_B$ is trivial.

On the other hand it is easy to see that if $b \in B_A^1$ then $1-\chi_\varepsilon(bh_t)-\chi_\varepsilon(wbh_t)$ is precisely the characteristic function of the compact interval $[H(wn)-\log\varepsilon, \log\varepsilon]$. This implies that $I_r(x,f,\varepsilon)$ is integrable (since $f$ is of compact support). It also implies that the integral of $I_r(x,f,\varepsilon)$ equals

$$(9.47)\qquad (\log\varepsilon)c_G\sum_{\mu\in A_r}\int_{N_A}\int_K f(k^{-1}n^{-1}\mu nk)\,H(wn)\,dn\,dk$$

$$(9.48)\qquad -\frac{c_G}{2}\sum_{\mu\in A_r}\int_{N_A}\int_K f(k^{-1}n^{-1}\mu nk)\,H(wn)\,dn\,dk\ .$$

Observe that the $\varepsilon$-dependence of $I_r(x,f,\varepsilon)$ has been isolated entirely in the term (9.47) and then in a quite transparent fashion. This is significant because $\mathrm{tr}(R_0^+(f))$ does *not* depend on $\varepsilon$. Therefore (9.47) will have to cancel out with another $\varepsilon$-term sooner or later. The term (9.48) on the other hand, represents half of the parabolic contribution of $K(x,x)$ to the trace formula. As such it should not be expected to cancel out.

Now we deal similarly with $I_s(x,f,\varepsilon)$. Since $1-\chi_\varepsilon(bh_t)$ is the characteristic function of $(-\infty, \log\varepsilon]$,

$$\int_X I_s(x,f,\varepsilon)\,dx = \sum_{\mu\in Z_Q} \int_{Z_\infty^+ B_Q\backslash G_A} \sum_{\substack{\nu\in N_Q\\ \nu\neq e}} f(x^{-1}\mu\nu x)(1-\chi_\varepsilon(x))\,dx$$

$$= c_G \sum_{\mu\in Z_Q} \int_{-\infty}^{\infty} e^{-2t} \int_K \sum_{\nu\neq e} f(k^{-1}h_t^{-1}\mu\nu h_t k)(1-\chi_e(h_t)dtdk$$

$$= c_G \sum_{\mu\in Z_Q} \int_K \int_{-\infty}^{\log\varepsilon} e^{-2t} \sum_{\nu\neq e} f(k^{-1}h_t^{-1}\mu\nu h_t k)\,dtdk\ .$$

But for large *negative* values of $t$ (and all but finitely many $\mu\in Z_Q$, $\nu\in N_Q$) $h_t^{-1}b^{-1}\mu\nu b h_t$ cannot lie within the support of $f$. Therefore $I_s(x,f,\varepsilon)$ is indeed integrable over $X$. Its integral is the value at zero of the entire function

$$I_\varepsilon(z) = c_G \sum_{\mu\in Z_Q} \int_K \int_{-\infty}^{\log\varepsilon} \sum_{\nu\neq e} e^{-2t(1+z)} f(k^{-1}h_t^{-1}\mu\nu h_t k)\,dtdk\ .$$

To write $I_\varepsilon(0)$ in a more convenient form we need to observe that the *only* orbit of $A$ acting in $N-\{e\}$ (by conjugation) is the orbit of

$$\begin{bmatrix}1 & 1\\ 0 & 1\end{bmatrix} = n_0\ .$$

The isotropy group of this matrix being $Z$ we may write

$$I_\varepsilon(z) = \sum_{\mu\in Z_Q} \tilde\theta(\mu,z,f)$$

(9.49)
$$-\sum_{\mu\in Z_Q} c_G \int_{Z_\infty^+ A_Q\backslash A_A} \int_K \Big(\sum_{\nu\in N_Q} f(k^{-1}a^{-1}\mu\nu a k)\left|\frac{a_1}{a_2}\right|^{-(1+z)} \chi_\varepsilon(a)\,dadk\Big)$$

where

(9.50)
$$\tilde\theta(\mu,z,f) = c_G \int_{Z_\infty^+ Z_Q\backslash A_A} \int_K f(k^{-1}a^{-1}\mu n_0 a k)\left|\frac{a_1}{a_2}\right|^{-(1+z)} dadk\ .$$

This expression for $I_\varepsilon(z)$ is desirable because it shoves the $\varepsilon$-dependence of $I_\varepsilon(z)$ entirely into (9.49). Consequently one knows *a priori* that (9.49) must either go to zero as $\varepsilon \to \infty$ or cancel with some contribution from $-K''(x, f, \varepsilon)$. Actually in this case an argument involving Poisson summation (similar to the argument used to prove Lemma 9.13) shows that (9.49) equals

$$(9.51) \qquad -\sum_{\mu \in Z_Q} c_G \int_{Z_\infty^+ A_Q \backslash A_A} \int_K \hat{f}(0, \mu, k) \left|\frac{a_1}{a_2}\right|^{-z} \chi_\varepsilon(a)\, da\, dk$$

$$(9.52) \qquad +\sum_{\mu \in Z_Q} c_G \int_{(Z_\infty^+ A_Q \backslash A_A)} \int_K f(\mu) \left|\frac{a_1}{a_1}\right|^{-(z+1)} \chi_\varepsilon(a)\, da\, dk$$

*plus an entire function of* $z$ *whose value at* $0$ *approaches zero as* $\varepsilon \to \infty$.

Now for $\mathrm{Re}(z) > 0$, the integral in (9.51) converges and equals

$$-(c_G)\, \frac{\varepsilon^{-z}}{z}\, (\log \varepsilon) \int_K \hat{f}(0, \mu, k)\, dk .$$

On the other hand, the integral in (9.52) converges *for* $\mathrm{Re}(z) > -1$ and in this range equals (some multiple of)

$$-\frac{(\varepsilon)^{-(1+z)}}{(1+z)}\, (\log \varepsilon)(c_G)$$

(which obviously approaches $0$ at $z = 0$ as $\varepsilon \to \infty$). Consequently we can conclude that:

(1) the integral defining $\tilde{\theta}(\mu, z, f)$ is absolutely convergent for $\mathrm{Re}(z) > 0$;

(2) $\tilde{\theta}(\mu, z, f)$ can be analytically continued to a meromorphic function on $\mathbb{C}$ whose only singularities are poles at $z = 0$ and $z = -1$;

(3) modulo a term which approaches $0$ as $\varepsilon \to \infty$,

$$I_{\varepsilon}(0) = \lim_{z \to 0} \sum_{\mu \in Z_Q} \left\{ \tilde{\theta}(\mu, z, f) - (c_G)(\log \varepsilon) \frac{\varepsilon^{-z}}{z} \int_K \hat{f}(0, \mu, k)\, dk \right\}$$

Summing up then, we have:

PROPOSITION 9.17 (Contribution of the second parabolic term to the Trace Formula).

$$\int_X \{I_r(x, f, \varepsilon) + I_s(x, f, \varepsilon)\}\, dx = \frac{c_G}{2} \sum_{\mu \in A_r} \int_{N_A} \int_K f(k^{-1}n^{-1}\mu nk) H(wn)\, dn dk$$

$$+ \sum_{\mu \in Z_Q} \lim_{z \to 0} \frac{d}{dz} \{z\tilde{\theta}(\mu, z, f)\}$$

$$+ (\log \varepsilon) c_G \sum_{\mu \in A_Q} \int_{N_A} \int_K f(k^{-1}n^{-1}\mu nk)\, dn dk\,.$$

*Proof.* Put together (3) above with (9.47) and (9.48). ⊐

As already remarked, since the last expression in Proposition 9.17 depends on $\varepsilon$, it must cancel out with some contribution from the third parabolic term $-K''(x, f, \varepsilon)$.

7. *The Third Parabolic Term*

To calculate

$$\int_X K''(x, f, \varepsilon)\, dx$$

let us recall the functions

$$E'_{\varepsilon}(x, \phi, z) = \sum_{\delta \in B_Q \backslash G_A} E^0(\delta x, \phi, z) \chi_{\varepsilon}(\delta x)$$

and

(9.53) $$K_1(z,f,x) = \frac{1}{4\pi}\sum_{i,j} R_{ij}(f,z)E(x,\phi_i,z)\overline{E(x,\phi,z)} .$$

Setting

(9.54) $$E''_\varepsilon(x,\phi,z) = E(x,\phi,z) - E'_\varepsilon(x,\phi,z)$$

one can define $K'_\varepsilon(z,f,x)$ and $K''_\varepsilon(z,f,x)$ by replacing $E$ in the expression for $K_1(z,f,x)$ above by $E'_\varepsilon$ and $E''_\varepsilon$. Then

(9.55) $$K_1(x,x) = \int_{-i\infty}^{i\infty} K_1(z,f,x)d|z| ,$$
$$K'(x,f,\varepsilon) = \int_{-i\infty}^{i\infty} K'_\varepsilon(z,f,x)d|z| ,$$

and

(9.56) $$K''(x,f,\varepsilon) = \int_{-i\infty}^{i\infty} \{K_1(z,f,x) - K'_\varepsilon(z,f,x)\}d|z| .$$

All these functions are slowly increasing on any Siegel set.

Recall also that for any $t > 0$,

$$S(t) = \{x \in G_A : H(x) > \log t\} .$$

With $c_1$ equal to the constant of Lemma 9, fix $t > c_1$.

Denote by $S(t)^*$ the projection of $S(t)$ onto $X$ and the closure of the complement of $S(t)^*$ in $X$ by $G(t)$. Certainly $G(t)$ is compact, so $K''(x,f,\varepsilon)$ is integrable on $G(t)$, and

$$\int_X K''(x,f,\varepsilon)dx = \lim_{t\to+\infty} \int_{G(t)} K''(x,f,\varepsilon)dx .$$

If we substitute the expression (9.56) for $K''(x,f,\varepsilon)$ and use the relation (9.54) we conclude that the integral over $X$ of $K''(x,f,\varepsilon)$ is

$$\lim_{t\to+\infty}\int_{-i\infty}^{i\infty}\int_{G(t)} K''_{\varepsilon}(z,f,x)\,dx\,d|z|$$

$$+\lim_{t\to+\infty}\int_{-i\infty}^{i\infty}\int_{G(t)}\frac{1}{4\pi}\sum_{i,j}R_{ij}(f,z)$$

$$\times\{E'_{\varepsilon}(x,\phi_i,z)E''_{\varepsilon}(x,\phi_j,z)+E''_{\varepsilon}(x,\phi_i,z)E'_{\varepsilon}(x,\phi_j,z)\}\,dx\,d|z|\ .$$

LEMMA 9.18. *For any* $i, j$,

$$\int_{G(t)} E'_{\varepsilon}(x,\phi_i,z)\overline{E''_{\varepsilon}(x,\phi_j,x)}\,dx = 0\ .$$

*Proof.* First note that since $E(x,\phi_j,z)$ is an automorphic form, the integrand above really is integrable over X. Its integral equals

$$\int_{Z_{\infty}^{+}B_{Q}\backslash G_{A}} E^{0}(x,\phi_i,z)\overline{E''_{\varepsilon}(x,\phi_j,z)}\chi_{\varepsilon}(x)\,dx$$

$$=\int_{Z_{\infty}^{+}N_{A}B_{Q}\backslash G_{A}} E^{0}(x,\phi_i,z)\ \chi_{\varepsilon}(x)\int_{N_{Q}\backslash N_{A}}\overline{E''_{\varepsilon}(nx,\phi_j,z)}\,dn\ \ dx\ .$$

But for $x \in S(\varepsilon)$, $E'_{\varepsilon}(x,\phi_j,z) = E^{0}(x,\phi_j,z)$. Therefore the constant term of $E''_{\varepsilon}(x,\phi_j,z)$ is zero. Similarly, using the assumption that $t > c_1$, on can show that the integral of $E'_{\varepsilon}(x,\phi_i,z)\overline{E''_{\varepsilon}(x,\phi_j,z)}$ over $S(t)^*$ also is zero, whence the lemma. □

From Lemma 9.18 it follows that to evaluate the contribution from $K''(x,f,\varepsilon)$ we need only evaluate

$$\lim_{t\to+\infty}\int_{-i\infty}^{i\infty}\int_{G(t)} K''_{\varepsilon}(z,f,x)\,dx\,d|z|$$

which in turn equals

$$(9.57)\qquad \frac{1}{4\pi}\sum_{i,j}\int_{-i\infty}^{i\infty} R_{ij}(z,f)\left\{\int_X E''_\varepsilon(x,\phi_i,z)\overline{E''_\varepsilon(x,\phi_j,z)}\,dx\right\} d|z|\,.$$

Fortunately, from the theory of Eisenstein series, we have:

LEMMA 9.19. *For any* $i,j$ *and* $z \in i\mathbb{R}^{\times}$,

$$\int_X E''_\varepsilon(x,\phi_i,z)\overline{E''_\varepsilon(x,\phi_j,z)}\,dx$$

*is the sum of*

$$2(\log\varepsilon)(\phi_i,\phi_j) - \frac{1}{2}\left\{(M(-z)\frac{d}{dz}(M(z))\phi_i,\phi_j) - (\phi_i, M(-z)\frac{d}{dz}(M(z))\phi_j)\right\}$$

*and*

$$\frac{1}{2z}\{\varepsilon^{2z}(\phi_i, M(z)\phi_j) - \varepsilon^{-2z}(M(z)\phi_i,\phi_j)\}$$

The proof of this result is to be found in Sections 4 and 6 of [Langlands 5]. Applying it to (9.57) yields at last:

PROPOSITION 9.20 (Contribution of the third parabolic term to the Trace Formula):

$$-\int_X K''(x,f,\varepsilon)\,dx = -\frac{\log\varepsilon}{2\pi}\int_{-i\infty}^{i\infty} \mathrm{tr}(R(f,z))\,d|z|$$

$$+\frac{1}{4\pi}\int_{-i\infty}^{i\infty} \mathrm{tr}\ M(-z)\frac{d}{dz}(M(z))R(f,z)\ \ d|z|$$

$$-\frac{1}{4}\,\mathrm{tr}\,\{M(0)R(f,0)\}$$

plus an expression which approaches $0$ as $\varepsilon\to\infty$.

REMARK 9.21. The "final form" of the trace formula is now close at hand. Indeed since $\varepsilon$ was assumed to be *arbitrarily* larger than $c_1$, the contributions from the elliptic, singular, or parabolic terms which depended on $\varepsilon$ but approached 0 as $\varepsilon$ approached $\infty$ can be (and have been) discarded. On the other hand, since $\operatorname{tr}(R_0^+(f))$ does *not* depend of $\varepsilon$, those $\varepsilon$-dependent terms which remain must (and do) cancel each other out. In particular, the first expression in Proposition 9.20 cancels with the $\varepsilon$-dependent contribution from the second parabolic term (Proposition 9.17). (This follows from substituting for $\operatorname{tr}(R(f,z))$ the integral of the kernel $P(f,z,x,x)$ described in the proof of Lemma 9.14.

## 8. *Final Form of the Trace Formula*

Putting together Remark 9.21 with Propositions 9.10, 9.16, 9.17, and 9.20, we conclude that the trace of $R_0^+(f)$ equals the sum of

$$\text{measure}\,(Z_\infty^+ G_Q \backslash G_A) \sum_{\mu \in Z_Q} f(\mu)\ ,$$

$$\sum_{\gamma \in \{G_e\}} \text{measure}\,(Z_\infty^+ G(\gamma)_Q \backslash G(\gamma)_A) \int_{G(\gamma)_A \backslash G_A} f(x^{-1}\gamma x)\,dx\ ,$$

$$-\frac{c_G}{2} \sum_{\mu \in A_r} \int_{N_A} \int_K f(k^{-1}n^{-1}\mu nk)\,H(wn)\,dn\,dk$$

$$\sum_{\mu \in Z_Q} \lim_{z \to 0} \frac{d}{dz} \{z\tilde{\theta}(\mu, z, f)\}\ ,$$

$$-\frac{1}{4\pi} \int_{-i\infty}^{i\infty} \operatorname{tr}\left\{M(-z)\frac{d}{dz}(M(z))\,R(f,z)\right\} d|z|$$

and

$$-\frac{1}{4} \operatorname{tr} M(0)\,R(f,0)\ .$$

In applying this formula it is often convenient to rewrite some of the terms above as "local integrals" as follows.

Recall that $H\left(\begin{bmatrix} a_1 & 0 \\ 0 & a_2 \end{bmatrix}\right) = \log \left|\frac{a_1}{a_2}\right|^{1/2}$. Therefore if we set

$$\lambda(n) = \left|\frac{a_1}{a_2}\right|^{1/2}$$

whenever $wn = n'\begin{pmatrix} a_1 & 0 \\ 0 & a_2 \end{pmatrix}k'$, the third term in the formula for the trace of $R_0^+(f)$ becomes

$$-\frac{c_G}{2}\sum_{\mu\in A_r}\int_{N_{\mathbf{A}}}\int_K f(k^{-1}n^{-1}\mu nk)\log\lambda(n)\,dndk$$

$$= -\frac{c_G}{2}\sum_{\mu\in A_r}\sum_{p(\text{prime})}\int_{N_{\mathbf{A}}}\int_K f(k^{-1}n^{-1}\mu nk)\log\lambda(n_p)\,dndk$$

$$= -\frac{c_G}{2}\sum_{\mu\in A_r}\sum_{p(\text{prime})}\int_{N_p}\int_{K_p} f_p(k_p^{-1}n_p^{-1}\mu n_pk_p)\log\lambda(n_p)\,dn_pdk_p \; x$$

$$\left(\prod_{w\neq p}\int_{N_w}\int_{K_w} f_w(k_w^{-1}n_w^{-1}\mu n_wk_w)\,dn_wdk_w\right)$$

since $\log\lambda(n) = \sum_p \log\lambda(n_p)$. (We are assuming that $f$ is of the form $\prod_p f_p$ with $f_\infty \in C_c^\infty(Z_\infty^+\backslash G_\infty)$, $f_p \in C_c^\infty(G_p)$ for all finite $p$, and $f_p$ the characteristic function of $K_p$ for almost every finite $p$.)

To deal with the fourth term, recall that

$$\tilde{\theta}(\mu, f, z) = c_G \int_{Z_\infty^+ Z_{\mathbf{Q}}\backslash A_{\mathbf{A}}}\int_K f(k^{-1}a^{-1}\mu n_0 ak)\left|\frac{a_1}{a_2}\right|^{-(1+z)} dadk\,,$$

where $n_0 = \begin{bmatrix} 1 & 1 \\ 0 & 1 \end{bmatrix}$ and $a = \begin{bmatrix} a_1 & 0 \\ 0 & a_2 \end{bmatrix}$. Now for each prime p, set

$$\tilde{\theta}(\mu, f_p, z) = \frac{1}{L(1+z, 1_p)} \int_{Z_p \backslash A_p} \int_{K_p} f_p(h_p^{-1} a_p^{-1} \mu n_0 a_p k_p) \left| \frac{(a_p)_1}{(a_p)_2} \right|^{-(1+z)} da_p dk_p ,$$

where $a_p = \begin{bmatrix} (a_p)_1 & 0 \\ 0 & (a_p)_2 \end{bmatrix}$ and $1_p$ is the trivial character of $Q_p^x$. Then

$$\lim_{z \to 0} \frac{d}{dz} \{z\tilde{\theta}(\mu, f, z)\} = \lim_{z \to 0} \frac{d}{dz} \left\{ c_G \prod_p zL(1+z, 1_p)\theta(\mu, f_p, z) \right\}.$$

But $L(1+z, 1) = \prod_p L(1+z, 1_p)$, so the Laurent expansion of $\prod_p L(1+z, 1_p)$ about $z = 0$ is

$$\frac{\lambda_{-1}}{s} + \lambda_0 + \cdots .$$

Therefore, the fourth term in the formula for the trace of $R_0^+(f)$ reads

$$\sum_{\mu \in Z_Q} c_G \left[ \lambda_0 \prod_p \theta(0, f_p, \mu) + \lambda_{-1} \left\{ \sum_p \theta'(0, f_p, \mu) \prod_{w \neq p} \theta(0, f_w, \mu) \right\} \right].$$

Summing up, we have:

THEOREM 9.22 (The Selberg Trace Formula). *The trace of* $R_0^+(f)$ *is given by the sum of the following terms*:

(i) $\quad \operatorname{meas}(Z_\infty^+ G_Q \backslash G_A) \sum_{\mu \in Z_Q} f(\mu)$ ,

(ii) $\quad \sum_{\gamma \in \{G_\ell\}} \operatorname{meas}(Z_\infty^+ G(\gamma)_Q \backslash G(\gamma)_A) \int_{G(\gamma)_A \backslash G_A} f(x^{-1} \gamma x)\, dx$ ,

(iii) $\quad -\frac{c_G}{2} \sum_{\mu \in A_r} \sum_p \left\{ D(\mu, f_p) \prod_{w \neq p} F_{f_w}^A(\mu) \right\}$ ,

(iv) $$c_G \sum_{\mu \in Z_Q} \lambda_0 \prod_p \theta(0,f_p,\mu) + \lambda_{-1} \sum_p \theta'(0,f_p,\mu) \prod_{w \neq p} \theta(0,f_w,\mu) \quad ,$$

(v) $$\frac{1}{4\pi}\int_{-i\infty}^{i\infty} \mathrm{tr}\left\{M(-z)\frac{d}{dz}(M(z))R(f,z)\right\} d|z| \ , \ \textit{and}$$

(vi) $$-\frac{1}{4}\,\mathrm{tr}(M(0)R(f,0)) \ .$$

*Here*

$$D(\mu, f_p) = \int_{N_p}\int_{K_p} f_p(k^{-1}n^{-1}\mu nk)\log\lambda(n)\,dndk \ ,$$

$$F_{f_w}^A(\mu) = \int_{N_w}\int_{K_w} f_w(k^{-1}n^{-1}\mu nk)\,dndk \ ,$$

$$\theta(z, f_w, \mu) = \frac{1}{L(1+z, 1_p)}\int_{Z_p\backslash A_p}\int_{K_p} f(k^{-1}a^{-1}\mu n_0 ak)\left|\frac{a_1}{a_2}\right|^{-(1+z)} dadk$$

*$\theta'(0, f_p, \mu)$ is the derivative of $\theta(z, f_p, \mu)$ at $0$, and $\lambda_{-1}$ and $\lambda_0$ are the first two terms of the Laurent expansion of $L(1+z, 1)$ about $z = 0$.*

REMARK 9.23. The contributions to the trace formula from the singular and elliptic conjugacy classes are reminiscent of the formulas which obtain in the case of *compact* quotient. Indeed, suppose $G$ is an algebraic reductive group which is defined over $\mathbb{Q}$ and such that

$$X = Z_\infty^+ G_{\mathbb{Q}} \backslash G_{\mathbb{A}}$$

is compact. (Examples of such groups arise from the multiplicative groups of the quaternion algebras considered in Section 10.) Let $R(g)$ as usual denote the natural representation of $G_{\mathbb{A}}$ in $L^2(X)$ and let $f(g)$ be (essentially) arbitrary in $C_c^\infty(Z_\infty^+\backslash G_A)$. Then the trace of $R(f)$ exists and equals

$$\sum_{\{\gamma\}} \text{meas}\,(Z_\infty^+ G(\gamma)_Q\backslash G(\gamma)_A) \int_{G(\gamma)_A\backslash G_A} f(x^{-1}(\gamma x)\,dx$$

the summation extending over a complete set of representatives of the conjugacy classes in $G_Q$.

The new features in the case of *non-compact* quotient (in particular GL(2)) are obviously the contributions from the *parabolic* conjugacy classes and Eisenstein series (which reflect the fact that some part of the spectrum, namely the continuous spectrum, has been cut away). These latter contributions (i.e. (iii)-(vi) in Theorem 9.22) are not as gruesome as they might appear to be (see Remark 9.26 below). Nevertheless it is a happy circumstance that for quite general f these terms actually vanish. Indeed, suppose $f_v$ is such that

$$\int_{N_v}\int_{K_v} f_v(k^{-1}ank)\,dn\,dk = 0$$

for all $a = \begin{bmatrix} a_1 & 0 \\ 0 & a_2 \end{bmatrix}$ in $A_v$. (A well-known example of such an $f_v$ is any $K_v$-finite matrix coefficient of a supercuspidal representation of $G_v$.) Then it can be shown that (in the terminology of Theorem 9.22)

$$F_{f_v}^A(\mu) = 0 = \theta(0, f_v, \mu)$$

for all $\mu$, and $\text{tr}(R(f,z)) = 0$. Consequently the trace formula simplifies considerably:

COROLLARY 9.24. *Suppose* $f = \Pi\, f_p$, *and suppose that for at least two primes* p,

$$\int_{N_p}\int_{K_p} f_p(k^{-1}ank)\,dn\,dk = 0\ .$$

*Then*

$$\operatorname{tr} R_0^+(f) = \operatorname{meas}(Z_\infty^+ G_Q \backslash G_Q) \sum f(\mu)$$

$$+ \sum_{\gamma \in \{G_e\}} \operatorname{meas}(Z_\infty^+ G(\gamma)_Q \backslash G(\gamma)_A) \int_{G(\gamma)_A \backslash G_A} f(x^{-1}\gamma x)\,dx \ .$$

This simple consequence of the trace formula will be seen to have powerful applications to number theory in Section 10.

REMARK 9.25 (Concerning an Explicit Trace Formula). One of the fundamental problems of harmonic analysis on groups is to find the Fourier transforms of certain naturally defined invariant distributions. Let us recall that a distribution $T$ is *invariant* on $G$ if $T(f^y) = T(f)$ for every $f \in C_c^\infty(G)$, and $y \in G$, where $f^y(x) = f(y^{-1}xy)$. The basic invariant distributions on $G$ are the characters of irreducible representations of $G$; these distributions are basic because (like $e^{ixy}$ for $\mathbb{R}$) they are *eigendistributions* for the center of the universal enveloping algebra of $G$.

The basic problem just alluded to amounts to finding a linear functional on the space of invariant eigendistributions which reproduces $T(f)$. For example, if $T$ is the distribution described by

$$R : f \to f(e) \ ,$$

then a solution to this problem is "simply" the Plancherel formula for $G$. On the other hand, if

$$T(f) = \operatorname{tr} R_0^+(f) \ ,$$

this problem amounts to finding an *explicit* form of the trace formula analogous to (9.16) in the case of compact quotient. To resolve it one needs to individually analyze the distributions described by (i)-(vi) in Theorem 9.22. (*WARNING*: although the *sum* of these terms defines an *invariant* distribution each term individually need *not* be invariant.)

It turns out that each of the terms (i), (ii), and (vi) *are* invariant but that (iii), (v) and (vi) are *not*. Of course the distributions (vi) and (vii) are already expressed in terms of characters (involving the representations $R(x,z)$) so we need only consider the remaining terms (i)-(iv). Of these, the simplest to deal with are (i) and (ii). In (iii) and (iv) one is reduced to considering the *local* distributions

$$f \to D(\mu, f_p) ,$$

$$f \to F^A_{f_p}(\mu)$$

and

$$f \to \theta(0, f_p, \mu) \quad \text{or} \quad \theta'(0, f_p, \mu) .$$

The first of these distributions is non-invariant but can certainly be "inverted" in the real case. Indeed if $p = \infty$, $\lambda(n) = (1+x^2)^{-\frac{1}{2}}$ if $n = \begin{bmatrix} 1 & x \\ 0 & 1 \end{bmatrix}$. In the p-adic case, $\lambda(n) = \max\{1, |x|^{-1}\}$. Although the requisite computations have yet to be done in this case they almost surely can be, given our current knowledge of $GL(2)$ over the p-adics.

The distribution $F^A_{f_p}(\mu)$ is invariant and well studied and its "Fourier transform" is concentrated entirely on principal series representations. The distributions involving $\theta(0, \mu, f_p)$, on the other hand, are considerably more difficult to deal with. In the generality of an arbitrary reductive group the investigation of almost all these distributions remains an essentially open area for research. Initial results in the rank one case have been obtained recently by [Sally-Warner] and [Arthur 2].

## FURTHER NOTES AND REFERENCES

As the name implies, the Selberg trace formula was introduced by [Selberg] in 1956. The first *complete* (published) proof however appeared only recently. In [Duflo-Labesse] a complete proof is given for $PGL(2)$. In [Jacquet-Langlands] a proof for $GL(2)$ more along the original lines of Selberg's approach is sketched. A complete proof for arbitrary reductive groups *of rank one* (along these same lines) is the subject matter of [Arthur].

We have included a complete proof for GL(2) because all the main ideas of the rank one proof already appear and then more clearly (for some people).

If the reader wishes to compare our Theorem 9.22 with the trace formula described on p. 516 of [Jacquet-Langlands] he should keep in mind that our term (iv) may be expressed as the sum of invariant and non-invariant expressions (which correspond respectively to Jacquet-Langlands' (vii) and (viii)). Furthermore, in Jacquet-Langlands' term (iv), there is a misprint: c should read c/2. Their terms (ii) and (iii) correspond to our (ii) but reflect the fact that they are working over an *arbitrary* global field. This generality requires little additional work. (Since elliptic conjugacy classes in $G_F$ are indexed by quadratic extensions of F the two terms for Jacquet-Langlands' elliptic contribution reflect the fact that (in the case of non-zero characteristic) some of these extensions are separable and others are not.)

Explicit forms of the trace formula in the case of *compact* quotient are described in [Gelfand-Graev-Pyatetskii-Shapiro]. In the case of *non-compact* quotients the complexity of such a formula has already been noted. Nevertheless for some applications a compromise is still sufficient for dramatic results. The case in point is the computation of the trace of the classical Hecke operators T(p) (acting on $S_k(\Gamma_0(N))$). The application of the trace formula to this problem was carried out by Selberg in his original paper and is described in representation theoretic language in [Duflo-Labesse]. The idea is basically the following.

Using harmonic analysis on GL(2) over local fields (in particular the reals) one can show that there is a function

$$f = \prod_{p \leq \infty} f_p$$

in $C_c^\infty(Z_\infty^+ \backslash G_A)$ such that $R_0^+(f)$ annihilates the orthocomplement of (the image of) $S_k(SL(2,Z))$ in $L^2(X)$ but on $S_k(SL(2,Z))$ coincides with the familiar action of $p^{k/2-1}T(p)$. (Cf. Lemma 3.7; the crucial part is the choice of $f_\infty$.) The point is that *for this choice of* f

$$\operatorname{tr} R_0^+(f) = \operatorname{tr}\left\{p^{\frac{k}{2}-1}\,T(p)\right\}$$

so the computation of the trace of $T(p)$ is reduced to "inverting" $\operatorname{tr} R_0^+(f)$ *for a very special* choice of f. The ensuing computations are by no means trivial but they are possible.

The result is the "classical formula of Eichler-Selberg" described on p. 283 of [Duflo-Labesse]. A special case of this formula, corresponding to $p = 1$, essentially gives the *dimension* of $S_k(\Gamma_0(N))$. Although dimension formulas of this type are entirely classical (by Riemann-Roch; see for example [Shimura 1]) the formulas for the trace of $T(p)$ represent a real triumph for Selberg's theory.

Further applications of explicit (or semi-explicit) forms of the trace formula are described or referred to in [Gelfand-Graev-Pyatetskii-Shapiro], [Duflo-Labesse], [Labesse], and [Jacquet-Langlands]. If one wants to use the trace formula to prove directly that a given representation of $GL(2, \mathbb{A})$ occurs in $L^2(X)$ (without examining its L-function) it seems (unfortunately) that one has to have still more explicit p-adic information than is at present available.

## §10. AUTOMORPHIC FORMS ON A QUATERNION ALGEBRA

Let F denote a global field and D a (division) quaternion algebra defined over F. Thus D is a central simple (division) algebra of rank 4 over F. The multiplicative group $G'$ of D is a reductive algebraic group defined over F and the homogeneous space

$$X' = Z_\infty^+ G'_F \backslash G'_{\mathbb{A}}$$

possesses a $G'_{\mathbb{A}}$-invariant measure.

By an *cusp form on* $G'_{\mathbb{A}}$ we shall understand *any* irreducible unitary representation of $G_{\mathbb{A}}$ which occurs in the decomposition of the natural representation $R'$ of $G_{\mathbb{A}}$ in $L^2(X')$. The purpose of this section then is to describe the proof and consequences of the following result:

THEOREM 10.1. *There is a correspondence which associates to each irreducible unitary representation* $\pi'$ *of* $G'_{\mathbb{A}}$ *an irreducible unitary representation* $\pi$ *of* $GL(2, \mathbb{A})$ *so that if* $\pi'$ *is a cusp form on* $G'_{\mathbb{A}}$ *then* $\pi$ *is a cusp form on* $GL(2, \mathbb{A})$ *(provided* $\pi'$ *is not one-dimensional).*

This theorem does for quaternion algebras what Theorem 7.11 does for quadratic extensions of F. Both theorems provide natural solutions to the problem (Problem A) of relating automorphic forms on some group $G'$ to automorphic forms on GL(2).

To establish the correspondence alluded to in the first part of Theorem 10.1 one applies the construction of Weil with V equal to D instead of a quadratic extension L (cf. Sections 7.A and B). To prove the second (and more difficult) part of the theorem one could follow the lead of Theorem 7.11 and analyze the L-function $L(s, \pi)$ attached to $\pi$. In this case

$$L(s,\pi) = L(s,\pi') \tag{10.1}$$

where $L(s,\pi')$ is the L-function naturally attached to $\pi'$ via the method of Tate alluded to in the "Notes and References" of Section 6. Since this L-function can be shown to be entire and to satisfy the requisite functional equation (again by Tate's method) Theorem 10.1'follows from Jacquet-Langlands' characterization of cusp forms on GL(2).

The proof just sketched (in particular the identity (10.1)) demonstrates the naturality of the correspondence $\pi' \to \pi$. Nevertheless in Subsection B below we present a different proof which seems to be the "best possible." This proof involves the Selberg trace formula and is natural in the sense that it establishes an essentially functional analytic result (the occurrence of $\pi$ in a certain $L^2$ space) using *only* functional analysis (not the intermediary device of L-functions). This proof also allows one to *significantly* strengthen Theorem 10.1 as follows:

THEOREM 10.2. *Let* S *denote the finite set of places* v *in* F *such that* $D_v = D\otimes_F F_v$ *is (still) a division algebra. Then the correspondence*

$$\pi' \to \pi = \otimes \pi_v$$

*is one-to-one onto the collection of cusp forms on* $GL(2, A(F))$ *with* $\pi_v$ *square-integrable* for each $v \in S$.

In Subsection C we shall explain some of the possible consequences of this remarkable theorem. To appreciate one particular application we have in mind it is helpful to understand the connection between automorphic forms on a quaternion algebra and the classical theory of cusp forms. The point is that thus far we have restricted attention to automorphic forms on GL(2) and hence ultimately to classical forms automorphic with respect to some arithmetic Fuchsian group of *congruence type*. By dealing with division algebras one is led to consider certain arithmetic groups with *compact* fundamental domain in $SL(2,\mathbb{R})$. Automorphic forms with respect to such $\Gamma$ do not possess familiar Fourier expansions (because $\Gamma$ has no unipotent elements) and consequently such forms have vigorously resisted explicit construction and systematic analysis.

Several preliminary results are collected in Subsection A below. In the Subsection (D) we sketch yet another proof of Theorem 10.1 (due to Shimizu). This proof directly constructs the space of (theta) functions on which $\pi$ acts. Therefore it is analogous to Shalika-Tanaka's treatment of Theorem 7.11. Its approach is desirable because it says something very definite about the representability of classical cusp forms by theta-series associated to quaternary quadratic forms (a long-standing problem).

## A. Preliminaries

Let D denote a division quaternion algebra defined over F and for each place v in F put

$$D_v = D \otimes_F F_v .$$

We say then that v is *ramified in* D if $D_v$ is a division algebra. Otherwise v is *unramified* and $D_v$ is naturally isomorphic to $M(2,F_v)$. (We also say that D *splits* (or does not split at v) if v is unramified (resp. ramified) in D.)

The number of F-places ramified in D is finite and even and the set of such places will be denoted by S. This set completely determines D, i.e. if there is given a set consisting of an even number of places of F, then there exists a unique quaternion algebra D over F such that S is precisely the set of places in F ramified in D. (Unique up to isomorphism.)

For each place v, put

$$G'_v = D_v^{\times} .$$

Following the theory for GL(2) one should attach to each local group $G'_v$ a *Hecke group algebra* $\mathcal{H}(G'_v)$.

If v is *unramified* in D there is an isomorphism

$$\theta_v : D'_v \to M(2,F_v) .$$

To fix such an isomorphism we first fix a *maximal order* $\mathfrak{D}$ in D. (An order in D is any subring which contains a basis for D over F and is finite as a $O_F$-module.) Letting $\mathfrak{D}_v$ be the $O_v$-module in $D_v$ generated by $\mathfrak{D}$ we fix $\theta_v$ such that $\theta_v(\mathfrak{D}_v) = M(2, O_v)$. With this choice of $\theta_v$ we may identify $G'_v$ with $GL(2, F_v)$ and define $\mathcal{H}(G'_v)$ as the inverse image (under $\theta_v$) of the Hecke algebra of $GL(2, F_v)$.

We define $K'_v$ to be the maximal compact subgroup of $G'_v$ such that $\theta_v(K'_v) = K_v = GL(2, O_v)$. Thus for unramified v the notion of an admissible representation of $\mathcal{H}(G'_v)$ is obvious.

If v is ramified in D then $\mathfrak{D}_v$ is the *unique maximal order in* $D_v$, namely $\{x \in D_v : |N(x)|_v \leq 1\}$. Thus

$$K_v = \{x \in G'_v : |N(x)|_v = 1\}$$

is the maximal compact subgroup of $G'_v$ and $\mathcal{H}(G'_v)$ is defined to be the algebra of locally constant compactly supported functions on $G'_v$. (If v is archimedean, $K_v$ is simply the norm one group of Hamilton's quaternion algebra, and $\mathcal{H}(G'_v)$ is the sum of the space of measures defined by infinitely differentiable compactly supported functions on $G'_v$ which are bi-$K'_v$-finite and the space of measures defined by matrix coefficients of finite dimensional representations of $K'_v$.)

An admissible representation $\pi'$ of $\mathcal{H}(G'_v)$ for v ramified has (once again) the obvious definition. For such an irreducible representation there exists an irreducible representation of $G'_v$ which "lifts" to $\pi'$ on $\mathcal{H}(G'_v)$. This representation of $G'_v$ will be admissible in the sense of Definition 4.9. Actually it will be *finite dimensional* since $G'_v$ is compact modulo its center. Conversely, every finite-dimensional continuous representation of $G_v$ is admissible.

As for the global theory, let

$$K' = \prod_v K'_v \ .$$

Let $\mathcal{H}(G'_{\mathbf{A}})$ denote the space spanned by all functions $f' = \Pi f'_v$ with $f_v \in \mathcal{H}(G'_v)$ and $f_v$ equal to the characteristic function of $K'_v$ for almost every v. Then a representation $\pi'$ of $\mathcal{H}(G'_{\mathbf{A}})$ in V is called *admissible* if:

(i) Every w in V is of the form $\sum_i \pi'(f_i)w_i$ with $f_i$ a *function* in $\mathcal{H}(G'_{\mathbf{A}})$;

(ii) $\pi'(\xi)$ is finite-dimensional if $\xi$ is an elementary idempotent of $\mathcal{H}(G'_{\mathbf{A}})$ (i.e. of the form $\Pi\xi_v$ with $\xi_v$ a linear combination of characters of irreducible representations of $K'_v$).

As for GL(2), a representation of $G'_{\mathbf{A}}$ will be called admissible if its restriction to K contains any irreducible of K at most finitely many times. It should not seem surprising that every *irreducible* such representation defines such a representation of $\mathcal{H}(G'_{\mathbf{A}})$ and is completely factorizable as

$$\bigotimes_v \pi'_v .$$

*Some Arithmetic*

In this paragraph we collect some facts which make it possible to relate the homogeneous space $Z^+_\infty G'_F \backslash G'_{\mathbf{A}}$ to the classical space

$$\Gamma \backslash \{\mathrm{Im}(z) > 0\} .$$

Here $\Gamma$ is some lattice subgroup of $SL(2,\mathbf{R})$. In particular we assume that F is a totally real number field and that D is *indefinite* over F, i.e. at least one infinite place of F is unramified in D. To simplify matters as much as possible we shall actually assume that $F = \mathbf{Q}$.

Much of the arithmetic quoted here has been developed in greater generality by Eichler. (See the references at the end of this section.)

Let $\mathfrak{D}$ denote *any* maximal order in D. Suppose $\mathfrak{A}$ is a finite Z-module such that $\mathbf{Q}\mathfrak{A} = D$. Then $\mathfrak{A}$ is a *right* (resp. *left*) $\mathfrak{D}$-ideal if $\mathfrak{A}\mathfrak{D} \subset \mathfrak{A}$ (resp. $\mathfrak{D}\mathfrak{A} \subset \mathfrak{A}$). Two right (resp. left) $\mathfrak{D}$-ideals $\mathfrak{A}$ and $\mathfrak{B}$ are called *equivalent* if $\delta\mathfrak{A} = \mathfrak{B}$ (resp. $\mathfrak{A}\delta = \mathfrak{B}$) for some $\delta \in D$. The number of equivalence classes of right $\mathfrak{D}$-ideals is finite and equal to the number of equivalence classes of left ideals. This number is independent of the choice of maximal ideal $\mathfrak{D}$ and (by analogy with the case of fields) is called the *class number of* D. *For maximal orders and indefinite* D *the class number is one.*

Now consider the group of ideles $G'_A$ of D. Its subgroup of x such that $x_p$ is a unit of $\mathfrak{D}_p$ for all finite p is precisely $G'_\infty \prod_{p<\infty} K'_p$. On the other hand, for *every* $x \in G'_A$, there is a (uniquely determined) right $\mathfrak{D}$-ideal $\mathfrak{A}$ such that $\mathfrak{A}_p = x_p\mathfrak{D}_p$. If we denote this right $\mathfrak{D}$-ideal $\mathfrak{A}$ by $\cap x_p\mathfrak{D}_p$ then *the mapping*

$$x \to \cap x_p\mathfrak{D}_p$$

*establishes a one-to-one correspondence between* $D^\times\backslash G'_A/G'_\infty \prod_{p<\infty} K'_p$ *and equivalence classes of right* $\mathfrak{D}$*-ideals.*

LEMMA 10.3. $G'_A = G'_Q(G'_\infty)^0 \prod_{p<\infty} K'_p$.

*Proof.* Since $G'_\infty = GL(2,\mathbb{R})$, $(G'_\infty)^0 = GL^+(2,\mathbb{R})$. From the remarks above (in particular the fact that the class number of D is one) it follows that $G'_A = G'_Q G'_\infty \prod K'_p$. Thus the lemma follows from the fact that maximal orders contain units of norm $-1$. □

REMARK. If F is any number field then

$$A(F)^\times = F^\times (F_\infty)^0 \left( \prod_{v \text{ finite}} O_v^\times \right)$$

if and only if the class number of F is one. That is, there is a one-to-one correspondence between equivalence classes of O-ideals (right or left since F is commutative) and the homogeneous space

$$F^\times\backslash A(F)^\times / F_\infty^0 \left( \prod_{v \text{ finite}} O_v^\times \right).$$

(Compare the "Notes and References" of Section 3.) So Lemma 10.3 simply generalizes this fact to non-abelian division algebras and its proof too follows the abelian case.

COROLLARY.

$$Z_\infty^+ G'_{\mathbb{Q}} \backslash G'_{\mathbb{A}} / \prod_{p<\infty} K'_p \cong \Gamma \backslash SL(2,\mathbb{R}) \tag{10.2}$$

*where*

$$\Gamma = G'_{\mathbb{Q}} \cap (G'_\infty)^+ \prod_{p<\infty} K'_p \,. \tag{10.3}$$

*Thus the norm one group of a maximal order of a quaternion division algebra provides a Fuchsian group with compact fundamental domain.*

The compactness of the quotient space

$$Z_\infty^+ G'_F \backslash G'_{\mathbb{A}}$$

is established, for example, in [Weil 4].

REMARK 10.4. The dense subspace of $K'$-finite $\mathfrak{z}'$-finite functions in $L^2(X')$ coincides with the space of functions $\phi(g)$ on $G'_{\mathbb{A}}$ which satisfy the following properties:

(i) $\phi(\gamma g) = \phi(g)$ for all $\gamma \in G'_F$;

(ii) $\phi(zg) = \phi(g)$ for all $z \in Z_\infty^+$;

(iii) $\phi(g)$ is right $K'$-finite;

(iv) as a function of $G'_\infty$ alone, $\phi$ is right $\mathfrak{z}'$-finite ($\mathfrak{z}'$ being the center of the universal enveloping algebra of the complexified Lie algebra of $G'_\infty$); and

(v) $\phi$ is "slowly increasing." We denote this space by $A(X')$. Since $X$ is compact, the continuity of $\phi(g)$ (together with the transforming property (ii) above) actually implies that Condition (v) is redundant. We include it to stress that $A(X')$ is precisely the analogue for $D$ of the space of automorphic forms for $GL(2)$.

In case D is an *indefinite* division algebra over $\mathbb{Q}$, $G'_\infty = GL(2,\mathbb{R})$. In this case Lemma 10.3 implies that special functions in $A(X')$ (namely those satisfying (3.6)(iii), (3.6)(iv) and $\phi(gk') = \phi(g)$ for all $k \in \prod_{p<\infty} K'_p$) correspond to functions $f(z)$ in $\{\mathrm{Im}(z) > 0\}$ satisfying the following properties;

(i) $f\left(\frac{az+b}{cz+d}\right) = (cz+d)^k f(z)$ for all $\begin{bmatrix} a & b \\ c & d \end{bmatrix} \in \Gamma = G'_{\mathbb{Q}} \cap GL^+(2,\mathbb{R}) \prod_{p<\infty} K'_p$; and

(ii) $f(z)$ is holomorphic.

Hence the terminology "automorphic form" for an element of $A(X')$ (or for an irreducible constituent of $R'_\psi$) is apt. Note that no cuspidal or boundedness condition is relevant since

$$\Gamma \backslash \{\mathrm{Im}(z) > 0\}$$

is compact. Note also that if $\phi$ is right SO(2)-invariant and $\phi$ satisfies $\Delta\phi = \frac{1-s^2}{4}\phi$ (in place of Conditions (3.6)(iii) and (iv)) then $f_\phi(z)$ is real analytic and a wave-form in the sense of Maass.

## B. Statement and Proof of the Fundamental Result

According to Theorem 7.6 one may use Weil's representation to associate to each irreducible unitary representation $\pi'$ of a local division quaternion algebra (over F) an irreducible unitary representation $\pi(\pi')$ of $GL(2,F)$. This representation $\pi$ will be a special representation if $\pi'$ is one-dimensional and super cuspidal otherwise. Furthermore all such representations of $GL(2,F)$ are so obtained.

THEOREM 10.5. *Let* D *devote a division quaternion algebra defined over the global field* F, S *the set of places of* F *ramified in* D, *and* G′ *the multiplicative group of* D. *To each irreducible unitary admissible representation* $\pi' = \bigotimes_v \pi'_v$ *of* $G'_{\mathbb{A}}$ *let* $\pi$ *denote the representation of* $G_{\mathbb{A}} = GL(2,\mathbb{A})$ *whose* v*-th component is equivalent to* $\pi'_v$ *if* $v \notin S$ *and equivalent to* $\pi_v(\pi'_v)$ *if* $v \in S$. *(Here* $\pi_v(\pi'_v)$ *denotes the irreducible component of the Weil representation* $r(D_v)$ *indexed by* $\pi'_v$.) *THEN*:

(i) $\pi = \otimes\pi_v$ *is a cusp form for* $G_A$ *if* $\pi'$ *is a (greater than one dimensional) cusp form for* $G'_A$;

(ii) *the mapping*

$$\pi' \to \pi ,$$

*restricted to the collection of (greater than one dimensional) cusp forms on* $G'_A$ *is* 1-1 *onto the collection of all (equivalence classes of) cusp forms* $\bigotimes_v \pi_v$ *on* $GL(2, A)$ *such that* $\pi_v$ *is square-integrable for each* $v \in S$.

REMARK. Suppose $\pi'$ has central character $\psi$. Then $\pi'$ is called a *cusp form* if it occurs in the natural representation $R'_\psi$ defined by right translation in the space $L^2(G'_F\backslash G'_A, \psi)$ consisting of measurable functions $\phi$ on $G'_F\backslash G'_A$ satisfying

$$\phi(zg') = \psi(z)\phi(g) \quad (\text{for all } z \in Z'_A = Z_A)$$

and

$$\int_{Z'_A G'_F\backslash G_A} |\phi(g)|^2\, dg < \infty .$$

Note that $\pi'$ will occur in the representation $R'$ acting in $L^2(X'))$ only if $\psi$ is trivial on $Z^+_\infty$. More generally (from the definition of $\pi(\pi')$ and the identity immediately following (7.16)) it follows that $\pi(\pi')$ again induces the character $\psi$ on $Z_A$. Therefore a more accurate statement of Theorem 10.5 is that

$$\pi' \to \pi$$

map the constituents of $R'_\psi$ one-to-one onto the constituents $\otimes\pi_v$ of $R_0^\psi$ such that $\pi_v$ is square-integrable for each $v \in S$.

One possible proof of Part (i) of Theorem 10.5 has already been alluded to. By analogy with Tate's theory for GL(1) of a *field* one introduces the *local zeta functions*

(10.4)
$$\zeta(f,\pi'_v,s) = \int_{G'_v = D^x_v} f(x)\pi'_v(x)\,|N(x)|_v^{s+1/2}\,d^x x$$

corresponding to each Schwartz-Bruhat function $f$ on $D_v$, each irreducible representation $\pi'_v$ of $D^x_v$, and complex number $s$. These integrals converge for $\mathrm{Re}(s)$ sufficiently large and extend to meromorphic functions in all of $\mathbb{C}$. Furthermore there exist Euler factors $L(\pi'_v, s)$ (of degree $\leq 1$) and scalar factors $\varepsilon(\pi'_v, s)$ such that

(10.5)
$$\frac{\zeta(f,\pi'_v,s)}{L(\pi'_v,s)}\,\varepsilon(\pi'_v,s) = -\frac{\zeta(\hat{f},\tilde{\pi}'_v,1-s)}{L(\tilde{\pi}'_v,1-s)}$$

(cf. (6.33)). When $\pi'_v$ is of dimension $> 1$ the factor $L(\pi'_v,s) = 1$. More generally,

(10.6)
$$L(\pi'_v,s) = L(\pi_v(\pi'_v),s)$$
$$L(\tilde{\pi}'_v,s) = L(\tilde{\pi}_v(\pi'_v),s)$$

and

(10.7)
$$\varepsilon(\pi'_v,s) = \varepsilon(\pi_v(\pi'_v),s)$$

for all $\pi'_v$. In this sense (among many others) the Weil correspondence $\pi'_v \to \pi_v(\pi'_v)$ is completely natural.

Now still following Tate, one can show that if $\pi'$ is a cusp form then the Euler products

$$L(s,\pi') = \prod_v L(s,\pi'_v)$$

and

$$L(s,\tilde{\pi}') = \prod_v L(s,\tilde{\pi}'_v)\ ,$$

initially defined for $\mathrm{Re}(s)$ sufficiently large, define meromorphic functions which are bounded in vertical strips of finite width and satisfy the functional equation

$$L(s,\pi') = \varepsilon(s,\pi')L(1-s,\pi')$$

if

$$\varepsilon(s,\pi') = \prod_v \varepsilon(s,\pi'_v) .$$

Furthermore these products define *entire* functions of $s$ if $\pi'$ is not of the form $g \to \omega(N(g))$ for some grossencharacter $\omega$ of $F$. But for each character $\chi_v$ of $F_v$, $\pi_v(\pi'_v)^{\sim} = \pi_v(\tilde{\pi}'_v)$ and $\pi_v(\pi'_v \otimes \chi_v) = \pi_v(\pi'_v) \otimes \chi_v$ (lifting $\chi_v$ to $D_v$ thru the norm map $N_v$). Consequently by (10.6), (10.7) and Part (ii) of Jacquet-Langlands' Theorem 6.18, it follows that $\pi = \otimes \pi_v$ must be a cusp form on $G_A$, whence *Part (i)* of Theorem 10.5.

To prove Part (ii) of Theorem 10.5 it seems (at present) to be necessary to use the Selberg trace formula. Actually the approach we have in mind makes it possible to prove Parts (i) and (ii) together in one fell swoop. This approach is based on the observation that the local correspondence

$$\pi'_v \to \pi_v$$

(for each ramified place $v$) is natural from yet another point of view, namely character theory.

More precisely, let $R$ denote a set of representatives $E_v$ of (equivalence classes of) quadratic extensions of $F_v$. For simplicity assume that $F_v$ has characteristic zero so that every such $E_v$ is separable. For each $E_v$ choose an imbedding of $E_v$ in $M(2,F_v)$ *and* $D_v$ and recall that by the theorem of Skolem-Noether this imbedding is uniquely determined up to inner automorphism.

Now the multiplicative group of each extension $E_v$ imbeds as a Cartan subgroup of $G_v(=GL(2,F_v))$ or $G'_v(=D_v^{\times})$. Thus $R$ also indexes the set of (equivalence classes of) Cartan subgroups so obtained. In $G'_v$ this set indexes a *complete* set of representatives of Cartan subgroups but in $G_v$ it indexes only the so-called elliptic Cartans. The point is that

$$\chi_{\pi_v}(b) = -\chi_{\pi'_v}(b) \tag{10.8}$$

whenever b is regular in $E_v$ (regular means that the roots of the equation $x^2 - tr(b)x + N(b) = 0$ are distinct). This miraculous fact comprises the content of Proposition 15.5 of [Jacquet-Langlands].

According to (10.8) then, the "characters" of the global representations $\pi$ and $\pi'$ should be the same. Before using this fact to prove Theorem 10.5 we need to quote without proof a somewhat technical but quite believable result.

Suppose G is any locally compact unimodular group and Z is a closed subgroup of its center. For each character $\psi$ of Z let $L^1(\psi)$ denote the algebra (under convolution product) of all measurable f on G which satisfy $f(zg) = \psi^{-1}(z)f(g)$ for all $z \in Z$ and whose absolute values are integrable on $Z\backslash G$. Let $S(G)$ denote some dense subalgebra of $L^1(\psi)$ which is closed under the operation $f(g) \to f^*(g) = \overline{f(g^{-1})}$.

LEMMA 10.6. *Suppose $\pi^1$ and $\pi^2$ are two unitary (not necessarily irreducible) representations of G which induce the character $\psi$ on Z. Suppose also that the operators*

$$\pi^i(f) = \int_{Z\backslash G} f(g)\pi^i(g)\,dg \qquad (i = 1,2)$$

*are Hilbert-Schmidt for all f in S(G). Then $\pi^1$ and $\pi^2$ are equivalent iff*

$$\text{(10.9)} \qquad \mathrm{tr}\,\pi^1(f * f^*) = \mathrm{tr}\,\pi^2(f * f^*)$$

*for every f in S(G).*

Note that since $\pi^i(f * f^*) = \pi^i(f)\pi^i(f)^*$, the operators $\pi^i(f * f^*)$ are of trace class by our assumption on $\pi^i(f)$. The lemma is believable because for an *irreducible* unitary representation $\pi$ the distribution $\pi(f)$ is known to determine $\pi$.

*Proof of Theorem 10.5*

Suppose first that $\pi'$ is an irreducible unitary representation of $G'_A$ of dimension greater than one. So as not to obscure the basic idea of the proof let us also assume that each $\pi'_v$ has dimension greater than one *for all* v. (This is actually an assumption on the dimensionality of $\pi'_v$ for v in S only since for the remaining places it can be shown that $\pi'_v$ is *infinite* dimensional as soon as $\otimes\pi'_v$ is greater than one dimensional.)

Now let $\pi = \pi(\pi')$ denote the corresponding representation of $G_A$ and $\psi$ its central character. What has to be shown is that $\pi'$ is equivalent to some subrepresentation of $R'_\psi$ if and only if $\pi(\pi')$ is equivalent to some subrepresentation of $R_0^\psi$. To prove this we shall apply the criterion of Lemma 10.6.

To decide which G should be used in Lemma 10.6 let us consider the decomposition

$$R_0^\psi = \oplus \pi^j .$$

Each irreducible constituent $\pi^j = \otimes\pi^j_v$ acts on a certain subspace $V^j$ of $L_0^2(\psi)$ so we let $V^j_v$ denote the space of $\pi^j_v$. Next we restrict our attention to the direct sum of those representations $\pi^i$ whose v-th components are equivalent to $\pi_v(\pi'_v)$ for each $v \in S$. We denote this subrepresentation of $R_0^\psi$ by $R_S$.

For each $v \in S$ we fix a $K_v$-finite unit vector $u_v$ in each $V^i_v$ which is the image (under $\pi'_v \to \pi_v(\pi'_v)$) of some fixed $K'_v$-finite unit vector $u'_v$ in the space of $\pi'_v$. The direct sum

$$M = \bigoplus_i \left\{ \bigotimes_{v \in S} u_v \right\} \otimes \left\{ \bigotimes_{v \notin S} V^i_v \right\},$$

summed over i indexing $\pi^i$ of the type just described will then be invariant and irreducible under the action of

$$G_S = \{g = (g_v) : g_v = 1 \quad \text{for all} \quad v \text{ in } S\} .$$

On the other hand we can consider the decomposition

$$R_{\psi} = \oplus (\pi')^j$$

and its natural subrepresentation $R'_S$. Replacing $\pi_v(\pi'_v)$ by $\pi'_v$ and $u_v$ by $u'_v$ we can define a subspace $M'$ of $L^2(X')$ analogous to the subspace M described above. The group

$$G'_S = \{g = (g'_v) : g'_v = 1 \quad \text{for all} \quad v \in S\}$$

now acts irreducibly on $M'$ but this group is naturally isomorphic to $G_S$ via the local isomorphisms $\theta_v$ described earlier.

Our claim is that *to prove Theorem 10.5 it suffices to prove that these two natural representations of* $G_S$ *(on* M *and* $M'$*) are equivalent.*

Indeed in $R'_S$ we are dealing with all irreducible unitary representations $\pi'$ of $G'_A$ which occur in $R'$ and have v-th component equal to $\pi'_v$ for each $v \in S$. The assertion that the corresponding representations $\pi(\pi')$ occur in $R_0^{\psi}$ is equivalent to the assertion that the direct sum of these representations is equivalent to $R_S$. But the v-th component of each summand of this direct sum is automatically equivalent to the v-th component of each summand of $R_S$ for each $v \in S$. So the last assertion simply says that the remaining components are equal, or, alternately, that the representations of $G_S$ in M and $M'$ are equivalent! The argument for the converse direction is entirely similar.

So letting $\tau$ and $\tau'$ denote the representations of $G_S$ in M and $M'$ it remains to prove that $\tau$ and $\tau'$ are equivalent. To do this we shall apply Lemma 10.6 with $G = G_S$, $Z = \prod_{v \notin S} Z_v$, $\psi = \prod_{v \notin S} \psi_v$, $\pi^1 = \tau$, $\pi^2 = \tau'$, and S(G) equal to the linear span of the space of functions on $G_S$ of the form

$$f(g) = \prod_{v \notin S} f_v(g_v)$$

where

(i) for all v,

$$f_v(z_v g_v) = \psi_v^{-1}(z_v) f_v(g_v), \quad z_v \in Z_v \ ,$$

(ii) for all v, $f_v$ is bi-$K_v$-finite and of compact support modulo $Z_v$;

(iii) if v is infinite, $f_v$ is infinitely differentiable;

(iv) if v is finite, $f_v$ is locally constant; and

(v) for almost all finite v, $f_v$ is zero outside $Z_v K_v$ and on $Z_v K_v$ is equal to $\omega_v^{-1}(\det g)$ where $\omega_v$ is unramified and such that $\omega_v^2 = \psi_v$.

According to Lemma 10.6 what has to be shown is that

$$\text{(10.10)} \qquad \operatorname{tr} r(f * f^*) = \operatorname{tr} r'(f * f^*)$$

for all f in S(G). To verify (10.10) we shall bring into play two natural representations whose traces are easily calculated (using the trace formula) and then relate these traces to those appearing in (10.10).

The regular representation relating to the right side of (10.10) is simply $R'_\psi$. Indeed for each v in S let $f'_v$ denote the function

$$f'_v(g) = d(\pi'_v)\, \overline{(\pi'_v(g)\, u'_v, u'_v)}$$

where $d(\pi'_v)$ is the formal degree of $\pi'_v$ with respect to (a certain) Haar measure on $G'_v$. (The formal degree with respect to $dg'_v$ is defined by the orthogonality relations

$$\int_{Z'_v\backslash G'_v} <\pi'_v(g'_v)u_1,\tilde{u}_1>\overline{<\pi'_v(g_v)u_2,\tilde{u}_2>}\, dg'_v = \frac{1}{d(\pi'_v)} <u_1,u_2><\tilde{u}_2,\tilde{u}_1>$$

for all $u_1, u_2, u_2, u_1$.) If $\Phi'(g)$ is the function on $G'_A$ defined by

$$\Phi'(g) = \left\{\prod_{v \in S} f'_v(g_v)\right\} f * f^*(g'_S)$$

it follows (from the orthogonality relations (10.11)!) that $R'_\psi(\Phi')$ coincides with $r'(f*f^*)$ on $M'$ and annihilates its orthocomplement. Consequently

$$\text{trace } R'_\psi(\Phi') = \text{trace } r'(f*f^*) \ . \tag{10.12}$$

Now let us consider the left side of (10.10). Recall that by our assumption on each $\pi'_v$, $v \in S$, the representation $\pi_v(\pi'_v)$ is super cuspidal for such $v$ (assuming $S$ contains no infinite places). Thus the matrix coefficients of $\pi_v(\pi'_v)$ are compactly supported modulo $Z_v$ and in particular satisfy orthogonality relations of type (10.11) (without the primes everywhere). The measure $dg_v$ on $Z_v\backslash G_v$ may be normalized so that the formal degree of each $\pi_v(\pi'_v)$ equals $d(\pi_v)$, and consequently we set

$$f_v(g_v) = d(\pi_v)\,\overline{(\pi_v(g)u_v, u_v)}$$

for each $\pi_v = \pi_v(\pi'_v)$, $v \in S$. Defining a function $\Phi(g)$ on $G_{\mathbf{A}}$ by

$$\Phi_f(g) = \left\{\prod_{v \in S} f_v(g_v)\right\} f*f^*(g_S)$$

if $g = (g_v)g_S$, the result once again is that (with this choice of $\Phi$)

$$\text{trace } R_0^\psi(\Phi) = \text{trace } r(f*f^*) \ . \tag{10.13}$$

Now according to (10.12) and (10.13) we can at last complete the proof of Theorem 10.5 by establishing that

$$\text{trace } R_0^\psi(\Phi_f) = \text{trace } R'_\psi(\Phi'_f)$$

for all $f \in S(G_S)$. Not by chance this is precisely the type of identity which the trace formula can establish. Indeed by a slight modification of Remark 9.23 (replacing $L^2(Z^+_\infty G'_F\backslash G'_{\mathbf{A}})$ by $L^2(G'_F\backslash G'_{\mathbf{A}},\psi)$) we have that

$$\text{tr } R'_\psi(\Phi'_f) = \sum_{\{\gamma\}} \text{meas}\,(Z'_{\mathbf{A}}G'(\gamma)_F G'(\gamma)_{\mathbf{A}}) \int_{G'(\gamma)_{\mathbf{A}}\backslash G'_{\mathbf{A}}} \Phi'(x^{-1}\gamma x)\,dx \tag{10.14}$$

the summation extending over a set of representatives of conjugacy classes in $Z'_F \backslash G'_F$. On the other hand, by Corollary 9.24, we have that

$$\operatorname{tr} R_0^{\psi}(\Phi_f) = \operatorname{meas}(Z_A G_F \backslash G_A)\Phi(e) \tag{10.15}$$
$$+ \sum_{\substack{\gamma \in \{G_e\} \\ (\text{modulo } Z_F)}} \operatorname{meas}(Z_A G(\gamma)_F \backslash G(\gamma)_A) \int_{G(\gamma)_A \backslash G_A} \Phi(x^{-1}\gamma x)\,dx$$

the hypothesis of the corollary being satisfied since $S$ contains at least two places (which for convenience we are assuming finite) and for these places the matrix coefficients of $\pi_v(\pi'_v)$ satisfy

$$\int_{N_v}\int_{K_v} f_v(k^{-1}\, ank)\,dn\,dk = 0 \,. \tag{10.16}$$

To complete the proof of Theorem 10.5 we shall exploit (10.8) and the discussion immediately surrounding it to show that (10.14) and (10.15) are equal.

Assuming for the sake of simplicity that $F$ has characteristic zero, each $\gamma$ in $G'_F = D^x$ lies in some separable quadratic extension of $F$ imbeddable in $D$. Thus the conjugacy classes in $Z'_F \backslash G'_F$ are indexed by the elements (identified modulo $Z'_F$) of equivalence classes of quadratic extensions $E$ of $F$ which are imbeddable in $D$. Such extensions are precisely those $E$ such that

$$E \otimes_F F_v \text{ is a field}, \quad v \in S \,, \tag{10.17}$$

and consequently (10.14) may be rewritten as

$$\operatorname{trace}(R'_{\psi}(\Phi'_f)) = \operatorname{meas}(Z'_A G'_F \backslash G'_A)\Phi'(e)$$
$$+ \frac{1}{2}\sum_{Q'} \sum_{\substack{\gamma \in Z'_F \backslash B_F \\ \gamma \notin Z_F}} \operatorname{meas}(Z'_A B_F \backslash B_A) \int_{B_A \backslash G'_A} \Phi'(x^{-1}\gamma x)\,dx \,,$$

where $Q'$ denotes a set of representatives of equivalence classes of quadratic extensions satisfying (10.17), $B_F = B_F(E)$ is the centralizer of $E$ in $G'_F$, and $B_A = B_A(E)$ is the centralizer of $E$ in $G'_A$. Keeping in mind the definition of $\Phi'_f$ we have finally that

$$\operatorname{trace} R'_\psi(\Phi'_f) = \operatorname{meas}(Z'_A G'_F \backslash G'_A)\left\{\prod_{v \in S} d(\pi'_v)\right\} f * f^*(e)$$

$$+\frac{1}{2}\sum_{Q'}\sum_{\substack{\gamma \in Z'_F\backslash B_F \\ \gamma \notin Z'_F}} \operatorname{meas}(Z'_A B_F \backslash B_A)\left\{\prod_{v \in S} \frac{\chi_{\pi'_v}(\gamma^{-1})}{\operatorname{meas}(Z'_v\backslash B_v)}\right\}\int_{B_S\backslash G'_S} f * f^*(x^{-1}\gamma x)\, dx$$

Now we deal similarly with (10.15). Letting $Q$ denote a set of representatives for the equivalence classes of *all* quadratic extensions $E$ of $F$, and defining $B_F = B_F(E)$ and $B_A = B_A(E)$ as above, we have that

$$\operatorname{trace} R_0^\psi(\Phi_f) = \operatorname{meas}(Z_A G_F \backslash G_A)\left\{\prod_{v \in S} d(\pi_v)\right\} f * f^*(e)$$

$$+\frac{1}{2}\sum_{Q}\sum_{\substack{\gamma \in Z_F\backslash B_F \\ \gamma \notin Z_F}} \operatorname{meas}(Z_A B_F \backslash B_A)\int_{B_A\backslash G_A} \Phi(x^{-1}\gamma x)\, dx\ .$$

We observe that

$$\int_{B_A\backslash G_A} \Phi(x^{-1}\gamma x)\, dx\ ,$$

is equal to the product of

$$\prod_{v \in S}\int_{B_v\backslash G_v} f_v(x_v^{-1}\gamma x_v)\, dx_v \tag{10.19}$$

and

$$\int_{B_S\backslash G_S} f * f^*(x^{-1}\gamma x)\, dx\ ,$$

where $f_v(g_v)$ is the matrix coefficient $(\pi_v(g_v)u_v, u_v)$ multiplied by the formal degree $d(\pi_v)$. So since $Z = Z'$, $G_S = G'_S$, and

$$\text{meas}(Z'_A G'_F \backslash G'_A) = \text{meas}(Z_A G_F \backslash G_A)$$

(provided the Haar measure chosen on $G'_A$ is its Tamagawa measure) it remains only to prove that

$$\int_{B_v \backslash G_v} f_v(x_v^{-1} \gamma x_v) dx_v = \frac{1}{\text{meas}(Z'_v \backslash B_v)} \chi_{\pi'_v}(\gamma^{-1}) , \tag{10.20}$$

if $\gamma$ belongs to a subfield of D and equals zero otherwise. But (10.20) is equivalent (*modulo (10.8)*) to the assertion that

$$\int_{Z_v \backslash G_v} (\pi_v(x_v^{-1} \gamma x_v) u_v, u_v) dx_v = d(\pi_v)^{-1} \chi_{\pi_v}(\gamma^{-1}) , \tag{10.21}$$

if $\gamma \in E$ with $E \otimes_F F_v$ a field and

$$\int_{Z_v \backslash G_v} (\pi_v(x_v^{-1} \gamma x_v) u_v, u_v) dx_v = 0 \tag{10.22}$$

otherwise. Both these properties of matrix coefficients of supercuspidal representations are well known (cf. Jacquet-Langlands, Section 7, in particular Proposition 7.5; (10.22) is essentially equivalent to the Condition (10.16)). Consequently the proof of Theorem 10.5 is finally complete.

REMARK 10.7. In the proof of Theorem 10.5 we (at times implicitly) assumed that:

(i) the set of ramified primes in D does not include infinite primes (i.e. real places, since no non-trivial division algebra is defined over $\mathbb{C}$); and

(ii) $\pi'_v$ has dimension greater than one for all ramified v so that $\pi_v(\pi'_v)$ is super cuspidal as opposed to special.

It should be stressed that these assumptions were made purely for convenience of exposition. If $\pi_v(\pi'_v)$ is *special* for v ramified in D of course one can no longer define $\Phi_f$ using a matrix coefficient of $\pi_v(\pi'_v)$ since such coefficients are not integrable let alone compactly supported modulo $Z_v$. However one can still find a compactly supported function $f_v$ whose Fourier transform (roughly speaking) is zero everywhere except at $\pi_v(\pi'_v)$ (where it is one). So using this $f_v$ to define $\Phi_f$ the trace formula is applicable and one still concludes that

$$\operatorname{tr} R^0_\psi(\Phi_f) = \operatorname{tr} \tau(f * f^*) .$$

In particular assumption (ii) may be dropped.

As for assumption (i), suppose that $v = \infty$ *does* ramify in D and that $\pi'_v(h)$ is of the form $N(h)^{-n/2}\rho_n(h)$. Then $\pi_v(\pi'_v)$ is equivalent to the discrete series representation $\sigma(\mu_1, \mu_2)$ of "lowest weight" $n+2$ (cf. (7.19)) and in this case, as has already been remarked in the "Notes and References" of the last section, there still exists a compactly supported function $f_v$ which does the job (cf. [Duflo-Labesse], Section II.4). Thus the proof of Theorem 10.5 is complete.

## C. Construction of Some Special Automorphic Forms in the Case of Compact Quotient

To describe at least one automorphic form $\pi'$ on a given quaternion algebra D defined over Q we shall construct an automorphic form $\pi$ on GL(2) of the type described in Theorem 10.5. This means that the p-th component of $\pi$ must be square-integrable for each p dividing the discriminant of D. To construct $\pi$ we shall use grossencharacters of a quadratic extension L of Q since this is the only construction of cusp forms thus far described. The delicate point will be to choose $\lambda$ so that $\pi(\lambda)$ satisfies the requisite properties.

Recall that $\pi(\lambda) = \otimes \pi_p$ is defined as follows if $\lambda = \prod_v \lambda_v$ is a grossen-character of L. In the terminology of Section 7,

$$\pi_p = \pi(\lambda_v, \lambda_{v'})$$

if p splits in L and both v and v′ divide p. On the other hand, if p does *not* split in L, i.e. *if* $L \otimes_{Q} Q_p$ *is a field*, then $\pi_p$ equals the irreducible component $\pi_p(\lambda_v)$ of the Weil representation attached to the extension $L \otimes_Q Q_p$.

Recall that $\pi_p(\lambda_v)$ is a principal series representation if $\lambda_v$ is trivial on the norm one group of $L_v$ and a super cuspidal representation otherwise. Therefore the p-th component of $\pi(\lambda)$ will be square integrable for each $p \in S$ if and only if

(10.23) $$L \otimes_Q Q_p \text{ is a field}$$

for all $p \in S$ *and* $\lambda_v$ does not factor through the norm map $N_v$ for any v dividing $p \in S$.

This discussion implies that we should consider only quadratic extensions E of Q which are subfields of D, and grossencharacters $\lambda = \Pi \lambda_v$ of E such that $\lambda_v$ is "suitable" in the above sense for all v lying above the primes of S. In particular we have:

THEOREM 10.8. *Let* L *denote a quadratic subfield of* D. *Then*:

(i) *To each "suitable" character* $\lambda$ *of the adele group of* $L^x$ *one can attach an irreducible admissible representation* $\pi'(\lambda)$ *of* $G'_A$, *the adele group of* $D^x$;

(ii) *If* $\lambda$ *is a grossencharacter of* L (*i.e. an automorphic form for* L) *then* $\pi'(\lambda)$ *is an automorphic form for* $G'_A$, *i.e.* $\pi'(\lambda)$ *occurs in some* $R'_\psi$.

*Proof.* Theorems 7.11, 10,5, and the above remarks. □

REMARK 10.9. Consider the right regular representation of SL(2, R) in $L^2(\Gamma \backslash SL(2,R))$ with Γ as in (10.3). Thus Γ is the norm one group of a maximal order of a division algebra D and

$$\Gamma \backslash SL(2,R) \cong Z_\infty^+ G'_Q \backslash G'_A \Big/ \prod_{p<\infty} K_p .$$

A natural problem is to decompose this representation into irreducibles.

Since $\Gamma\backslash SL(2,\mathbf{R})$ is compact we know *a priori* that the decomposition is discrete. Furthermore each irreducible $\pi$ of $SL(2,\mathbf{R})$ occurs at most finitely many times and with multiplicity equal to the dimension of the corresponding space of automorphic forms of type $\pi$. (If $\pi = \pi_k^+$, $f(z)$ is holomorphic in $\mathrm{Im}(z) > 0$ and of weight $k$ with respect to $\Gamma$; if $\pi = \pi_s^+$ or $\pi_s^c$, then $f(z)$ is *invariant* with respect to $\Gamma$ and satisfies

$$\Delta f = -y^2\left(\frac{\partial^2}{\partial x^2} + \frac{\partial^2}{\partial y^2}\right) f = \frac{1-s^2}{4} f.)$$

Now the multiplicity with which $\pi_k^+$ occurs may be computed using the Selberg trace for compact quotient as explained in Section 9.A. Although this method fails to give a multiplicity formula for $\pi_s^+$ or $\pi_s^c$ we do know that such representations occur. Indeed since $L^2(\Gamma\backslash SL(2,\mathbf{R})/SO(2)) = L^2(\Gamma\backslash\{\mathrm{Im}(z) > 0\})$ is obviously infinite-dimensional it follows that *infinitely* many class 1 representations of $SL(2,\mathbf{R})$ occur in $L^2(\Gamma\backslash SL(2,\mathbf{R}))$. (Cf. the proof of Theorem 2.11.) The problem is which ones?

In the case of non-compact quotient we constructed examples of real analytic cusp forms using Theorem 7.11. Since Theorem 10.8 is an analogue of Theorem 7.11 (replacing $GL(2,\mathbf{A})$ by $G'_{\mathbf{A}}$) one should be able to use it to construct similar examples in the case of compact quotient. To explain this let us consider at least one specific example.

To simplify matters fix $D$ to be the unique (indefinite) division quaternion algebra defined over $\mathbf{Q}$ which splits at all primes except $p = 2$ and $3$. The quadratic extension

$$L = \mathbf{Q}(\sqrt{2})$$

of $\mathbf{Q}$ satisfies (10.23) at least for $p = 2$ and $3$ and consequently imbeds as a maximal subfield of $D$. (It is perhaps the "simplest" such subfield since $L \otimes_{\mathbf{Q}} \mathbf{Q}_p$ is ramified only when $p = 2$.)

According to Theorem 10.8 we should pick a grossencharacter $\Pi\lambda_v$ of $L$ such that $\lambda_v$ is "suitable" for $p = 2$ and $3$ and (say) unramified for all remaining $p$. Note that by our choice of $L$ the infinite prime $p = \infty$ splits so $\lambda_{v_\infty}$ and $\lambda'_{v_\infty}$ determine a continuous series representation

$\pi(s_1, s_2)$ of $GL(2,\mathbb{R})$. Consequently the construction of such grossen-characters *is* tantamount to the construction of examples of real analytic cusp forms for compact quotient. However, rather than pursue this point, we prefer to describe some related questions motivated by the remarks below.

REMARK. For GL(2) there is the *global* "multiplicity one" result of Theorem 5.7 whose proof uses Whittaker models (hence Fourier expansions of cusp forms along $\left\{\begin{bmatrix}1 & x\\ 0 & 1\end{bmatrix}\right\}$). Since this proof obviously does *not* extend to *division* quaternion algebras it is of interest to ask whether multiplicity one itself still holds.

THEOREM 10.10. *In the decomposition of* $R'_\psi$ *an irreducible representation of* $G'_{\mathbb{A}}$ *occurs at most once.*

*Proof.* If $\pi'$ occurs twice then the corresponding $\pi(\pi')$ would have to occur twice in $R_0^\psi$ contradicting multiplicity one for GL(2). Indeed *Theorem 10.5 establishes a one-to-one correspondence between (most of) the constituents of* $R'_\psi$ *and certain constituents of* $R_0^\psi$. *In particular it establishes an isomorphism between the subspaces of* $L^2$ *in which these representations act.*

REMARK 10.11. There is an important generalization of Theorem 10.8 which must be true but thus far seems too difficult to prove.

Suppose $D$ is a division algebra of *arbitrary* rank $(n^2)$ over $F$. Then one can define $G'_{\mathbb{A}}$ and $K'$ and $G'_F$ as in the case of rank 4. (Indeed $D_v$ will now be isomorphic to $GL(m_v, H_v)$ where $H_v$ is a division algebra of rank $n^2/m_v^2$ over $F_v$ and $v$ will be said to *ramify* in $D$ only if $m_v < n$.) The natural conjecture to make is:

CONJECTURE 10.12. Suppose $L$ is a maximal subfield of $D$ and $\psi$ is a character of $L_{\mathbb{A}}^{x}$. Then one may attach to $\psi$ an irreducible admissible representation of $G'_{\mathbb{A}}$. If $\psi$ is a *grossencharacter* and "suitable" then $\pi'(\psi)$ occurs in the space of automorphic forms for $G'_{\mathbb{A}}$.

How might one go about proving this conjecture? Unfortunately the proof of Theorem 10.8 does not go through since one knows little about L-functions for GL(n). (In particular one does not have a "Converse" to Hecke's theory at the present time.) An alternate approach might be to apply the trace formula directly to $G'$ without recourse to GL(n). Since this approach seems promising let us sketch what it already involves when $n = 2$.

According to Selberg's trace formula

$$\operatorname{tr}(R'_{\psi}(f)) = \int_{Z_A G'_F \backslash G'_A} \sum_{\gamma \in Z_F \backslash G'_F} \operatorname{tr}(f(x^{-1}\gamma x)dx$$

$$(10.24) \quad = \sum_{\gamma \in \{Z_F \backslash G'_F\}} \operatorname{meas}(Z_A G'(\gamma)_F \backslash G'(\gamma)_A) \int_{G'(\gamma)_A \backslash G'_A} \operatorname{tr}(f(x^{-1}\gamma x))dx$$

$$= \sum m_{\pi'} \operatorname{tr} \pi'(f)$$

if $R'_{\psi} = \oplus (m_{\pi'})\pi'$ and $f$ is a nice function on $G'_A$ (with values in the space of operators on some $K'$-module $\Lambda$; this extension to operator valued functions $f$ is straightforward).

Quite simply the idea is to pick $f = \Pi f_v$ so as to isolate $\pi'(\psi)$ in (10.24) in much the same way one isolates the multiplicity of a discrete series representation at infinity by choosing $f_\infty$ to be a matrix coefficient for this representation.

If $v$ ramifies in $D$ there is no problem since any matrix coefficient of $\pi'_v(\psi)$ will do the job. Now what happens at the remaining places?

At every unramified place *which does not split in* L, $\pi'_v(\psi)$ is either supercuspidal or a principal series representation of the form $\pi(\delta_v, \delta_v\eta_v)$ (with $\eta_v$ equal to the character of order two of $F_v^x$ associated to $L_v$). If $\pi'_v(\psi)$ is supercuspidal again there is no problem but of course this situation arises only finitely often. For the remaining non-splitting places, $\pi_v(\psi)$ is the extension to GL(2) of the (reducible) principal series representation of SL(2) induced from the character $\eta_v$ and this representation (like all induced representations) can not easily be isolated by Fourier transform.

What hope there is for completing this approach stems from the fact that each irreducible piece of the restriction of $\pi(\delta_v, \delta_v \eta_v)$ to SL(2) may be viewed a "limit" of discrete series representations. At the infinite prime this concept of "limit" is more than mere formal analogy so the guess is that these representations *do* distinguish themselves among principal series representations (vis-à-vis the trace formula).

Supposing for the moment that for all non-splitting v we *could* find $f_v$ so that

$$\int_{G(\gamma)_v \backslash G_v} f_v(x^{-1}\gamma x)\,dx = 0$$

whenever $\gamma$ lies in some quadratic extension of F which splits at v (the matrix coefficients of supercuspidal representations of $GL(2, F_v)$ have already been remarked to provide such functions) it would then follow that all conjugacy classes in $Z_F \backslash G_F'$ not indexed by elements of the extension L (describing $\pi(\psi)$) would contribute nothing to the trace in (10.24). Indeed quadratic extensions of F which don't split at any place where L doesn't split must equal L. In short we would have a formula for $\Sigma\, m_{\pi'} \operatorname{tr}\pi'(f)$ (summed over all $\pi'$ occurring in $R_\psi'$) which involved only the extension L.

Unfortunately the derivation of such a formula must await further developments in harmonic analysis.

## D. Theta Series Attached to Quaternary Quadratic Forms

In this subsection we describe how a special case of Theorem 10.5 may be applied to the so-called "basis problem" in the classical theory of automorphic forms.

Recall that theta-series attached to positive-definite quadratic forms define automorphic forms for $\Gamma_0(N)$. (Cf. Example B(iv) of Section 1.)

Roughly speaking, the "basis problem" asks whether all elements of $S_k(\Gamma_0(N),\psi)$ may be expressed as linear combinations of such theta-series in particular theta-series attached to the norm form of a (definite) division quaternion algebra. In representation theoretic language this amounts to the explicit construction of the space of $\pi = \otimes \pi_p$ in $L_0^2(G_Q\backslash G_A,\psi)$ given that $\pi$ corresponds to $\pi'$ through Theorem 10.5.

Our exposition here follows [Shimizu]. The work involved is analogous to the construction of Shalika-Tanaka described in Section 7C and the analogy is especially strong as regards the use of Weil's representation. In the present context this representation is used to construct (generalized) theta-functions attached to automorphic forms on division *quaternion algebras* as opposed to quadratic extensions.

The applications we have in mind deal with cusp forms of weight $k$ and character $\psi$ for $\Gamma_0(N)$ *with* $N$ *square free*. Thus we assume that $F = Q$ and that $D$ is definite and of discriminant $N = p_1 \cdots p_r$.

1. *Weil Representations and Theta Series*

The global Weil representation $r(s)$ of $SL(2,A)$ attached to $D$ is defined locally in terms of the completions $D_p$.

For each $p$ let $r_p$ denote the (local) Weil representation corresponding to $D_p = D \otimes_Q Q_p$ and its norm form. If $p \in S$, $D_p$ is the unique division quaternion algebra defined over $Q_p$, and $r_p$ is as described in Section 7A ((7.4)-(7.6)). On the other hand, if $p \notin S$, $D_p$ is identified with the vector space $V = M(2,Q_p)$ whose norm form is the usual determinant. In this case the (Weil) representation for $SL(2,Q_p)$ is described by (7.4)-(7.6) with $\omega(a) \equiv 1$, $\gamma = 1$, and

$$x^\sigma = \begin{bmatrix} x_1 & x_2 \\ x_3 & x_4 \end{bmatrix}^\sigma = \begin{bmatrix} x_4 & -x_2 \\ -x_3 & x_1 \end{bmatrix}.$$

Let $S_0(D_A)$ be the space spanned by all elements $\prod_p \Phi_p$ where for almost all $p$, $\Phi_p$ is the characteristic function of $\mathfrak{D}_p$. Since $r_p(s_p)(s_p \in SL(2,O_p))$ fixes the characteristic function of $\mathfrak{D}_p$ for almost all $p$ the formula

$$r(s)\Phi = \prod_p r(s_p)\Phi_p$$

is meaningful and defines the so-called *global Weil representation of* $SL(2,A)$ *attached to* $D$. This representation extends to $S(D_A)$ in such a way that the mapping $(s,\Phi) \to r(s)\Phi$ from $SL(2,A) \times S(L_A)$ into $S(D_A)$ is continuous. It also extends to a unitary representation in $L^2(D_A)$ such that (7.36) and (7.37) hold with $\gamma = 1$.

The space of generalized theta functions corresponding to the global representation $r(s)$ consists of functions

$$\Theta(r(s)\Phi) = \sum_{\xi \in D} (r(s)\Phi(\xi) = \Theta_\Phi(s) \tag{10.25}$$

(cf. (7.38) and (7.39)). The series in (10.27) converges uniformly and the function of $s$ it defines is continuous and left invariant for $SL(2,F)$. *To construct a subspace of* $L_0^2(G_A,\psi)$ *which realizes the representation* $\pi(\pi')$ *we decompose this space of theta-functions according to the irreducible representations* $\pi'$, i.e. we project $\Theta_\Phi(s)$ onto a cusp form which transforms under $G'_A$ "according to $\pi'$" (analogous to the projection (7.40)).

2. *Decomposition of the Weil Representation*

To decompose $r(s)$ we introduce certain *spherical functions* attached to $\pi'$. By $H^1$ we denote the norm one group of $D$ and then we put

$$K_p^1 = K_p' \cap H_p^1$$

and

$$K^1 = K' \cap H_A^1 = \Pi K_p'$$

If $d = \otimes d_p$ is any irreducible representation of $K^1$ *which occurs in the restriction of* $\pi'$ *to* $K^1$ we let $L(d)$ denote the space of all $\phi$ in the space of $\pi'$ such that

(10.26) $$\int_{K^1} \chi_d(k_1^{-1})\, \pi'(k_1)\phi dk_1 = \phi .$$

(Here $\chi_d(k_1) = (\dim d)$ trace $(d(k_1))$.)

Since $L(d)$ is finite-dimensional we can define the spherical function $\omega_d$ *of* $d$ *with respect to* $\pi'$ as

$$\omega_d(g) = \text{trace}\,(P(d)\pi'(g))$$

where $P(d)$ denotes projection onto $L(d)$. From this definition it follows that

$$\omega_d(g) = \sum_{i=1}^{N} (\pi(g)\phi_i, \phi_i)$$

if $\{\phi_1, \cdots, \phi_N\}$ is an orthonormal basis of $L(d)$. For the application we have in mind, the multiplicity of $d$ in $L(d)$ will be one, and in this case we have

(10.27) $$\omega_d(g) = \frac{\dim(d)}{\phi(g_0)} \int_{K^1} \phi(g_0 k_1 g k_1^{-1})\, dk_1$$

for any non-zero $\phi$ in $L(d)$ and any $g_0$ in $G'_A$ such that $\phi(g_0) \neq 0$. Since $\pi' = \otimes \pi'_v$, we always have

(10.28) $$\omega_d(g) = \prod_p \omega_{d_v}(g_v) .$$

Now we can describe the projection (or "component") of $\Theta_\Phi(s)$ which "transforms according to $\pi'$. First we restrict attention to $\Phi = \Pi\Phi_p$ such that $\Phi(k_1 x k_1^{-1}) = \Phi(x)$ for all $k \in K^1$,

(10.29) $$\int_{K^1} \chi_d(k_1)\Phi(k_1 x) dk_1 = \Phi(x) ,$$

and

(10.30) $$\Phi_\infty(x) = e^{-2\pi N(x)} P(N(x)) \operatorname{trace} \rho_n(x)$$

$P$ being some polynomial. For such $\Phi$, and $s$ in

$$GL(2,A)^+ = \{s \in GL(2,A) : \det(s) = N(h),\ h \in D_A^\times\}$$

we put

$\Theta_\Phi(d,\pi')(s)$

(10.31)
$$= |\det(s)|_A \sum_{i=1}^{N} \int_{Z'_A G'_Q \backslash G'_A} \overline{\phi_i(g)} \int_{H^1_Q \backslash H^1_A} \phi_i(nhg) \sum_{\xi \in D} r(s_1)\Phi^g(\xi nh))\, dndg$$

where $N(h) = \det(s)$, $\Phi^g(x) = \Phi(g^{-1}xg)$, and $s = \begin{bmatrix} \det(s) & 0 \\ 0 & 1 \end{bmatrix} s_1$ with $s_1 \in SL(2,A)$.

In case $d$ has multiplicity one in $L(d)$, (10.31) should simply read

(10.32) $$\Theta_\Phi(d,\pi')(s) = \frac{\dim(d)|\det(s)|}{\phi(g_0)} \int_{H^1_Q \backslash H^1_A} \phi(nhg_0)\Big(\sum_{\xi \in D} r(s_1)\Phi^{g_0}(\xi nh)\Big)\, dn$$

where $\phi \in L(d)$ and $\phi(g_0) \neq 0$. Note that in this case $\theta_\Phi(d,\pi')$ is defined completely analogously to the theta function $\theta_\Phi(\lambda)$ described by (7.40) (and attached to the "spherical function" $\lambda$). In any case $\Theta_\Phi(d,\pi')$ may be extended to a function on $GL(2,A)$ which is left invariant by $GL(2,Q)$ and an element of $A_0(\psi)$.

More precisely, let $\varepsilon_-$ denote the element of $GL(2,A)$ whose infinite component is $\begin{bmatrix} -1 & 0 \\ 0 & 1 \end{bmatrix}$ and whose remaining components are trivial. Let $V^*(\pi')$ be the subspace of $A_0(\psi)$ spanned by $\Theta_\Phi(d,\pi')$ (for all $\Phi$ of the type described above) and their translates by $\varepsilon_-$. *Then* $V^*(\pi')$ *is invariant for the natural action of the Hecke algebra* $\mathcal{H}(G_A)$ *and equivalent to (i.e. realizes)* $\pi(\pi')$.

To prove this it is convenient to analyze the first Fourier coefficients

$$W_\Theta(g) = \int_{Q\backslash A} \Theta\left(\begin{bmatrix} 1 & x \\ 0 & 1 \end{bmatrix} g\right) \overline{\tau(x)}\, dx$$

of each $\Theta$ in $V^*(\pi')$. Indeed the map

$$\Theta \to W_\Theta$$

commutes with the right action of $G_A$ so it suffices to show that this image space (call it $W^*(\pi')$) is equivalent to $\pi(\pi')$.

Now if $\Theta = \Theta_\Phi$,

$$W_\Theta(g) = |\det g|_A \int_{H_A^1} \omega_d(nh) r(g_1) \Phi(nh)\, dn$$

if $g = \begin{bmatrix} \det(g_1) & 0 \\ 0 & 1 \end{bmatrix} g_1$ belongs to $GL(2, A)^+$, and equals zero otherwise. But by (10.28), $\omega_d(g) = \Pi \omega_{d_p}(g_p)$, so since

$$\Phi = \Pi \Phi_p, \text{ and } r(x) = \otimes r_p(s)$$

matters reduce to analyzing the local functions

$$W_{\Theta_v}(g_v) = |\det g_v| \int_{K_v^p} \omega_{d_p}(nh) r_p(g_1^v) \Phi_p(nh)\, dn$$

on $GL(2, Q_p)$. These functions in turn are shown to generate the Whittaker spaces of $\pi_p(\pi_p')$ and therefore the desired conclusion follows.

3. *Application to the Basis Problem*

In this section we fix k and restrict our attention to representations $\pi'$ of $G_A'$ satisfying the following conditions:

(i) $\pi'$ has central character $\psi$;

(ii) $\pi'$ is a constituent of $R_\psi'$;

(iii) $\pi'_\infty(h) = N(h)^{-(k-2)/2}\rho_{k-2}(h)$ (compare the notation of (7.18);

and

(iv) the restriction of $\pi'_p$ to $K'_p$ contains the identity representation for all finite p.

Note that condition (ii) implies that for $p \in S$ we have $\pi'_p = \chi_p \cdot N$ with $\chi_p$ an unramified character of $Q_p^x$. In particular, $\pi'_p$ is one-dimensional for all $p \in S$. The representations $\pi(\pi') = \otimes \pi_p$ corresponding to these $\pi'$ then satisfy the following properties:

(i) $\pi(\pi')$ has central character $\psi$;

(ii) $\pi(\pi')$ is a constituent of $R_0^\psi$ (by Part (i) of Theorem 10.5);

(iii) $\pi_\infty = \sigma(\mu_1, \mu_2)$ where

$$\mu_1(a) = |a|^{r+k-3/2}$$

and

$$\mu_2(a) = |a|^{r-1/2}[\mathrm{sgn}(a)]^{k-2}$$

(cf. (7.19) and the notation of Section 4A);

and

(iv) the restriction of $\pi_p$ to $K_p(1) = \left\{\begin{bmatrix} a & b \\ c & d \end{bmatrix} \in GL(2,O_p) : c \equiv O(p)\right\}$ contains the identity representation.

(This follows from Theorem 7.6(ii) and table (4.20).)

Now we apply the construction of the last paragraph to these special $\pi'$ and a special choice of d, namely

$$d = d_\infty \otimes (\otimes d_p)$$

where $d_\infty$ is equivalent to $\rho_{k-2}$ (identifying $K^1_\infty$ with SU(2)) and $d_p$ is the identity representation. For this d matters simplify considerably since d has multiplicity one in L(d) (and therefore (10.32) is relevant).

In particular we may introduce the theta-functions

$$\theta_{i,j}(g) \quad (i, j = 1, \cdots, n)$$

as follows. Let H′ denote the space of $\phi$ in the space of any $\pi'$ right invariant for $\prod_{p<\infty} K'_p$. By our assumptions above this space is isomorphic to the space of all functions on $G'_{Q}\backslash G'_{A}$ taking values in the space U of matrix coefficients of $\pi'_\infty$ satisfying

$$\phi(hgk) = \left(\bigoplus_{\dim \pi'_\infty} \pi'_\infty\right)(g^{-1})\phi(h)$$

for all $h \in G'_A$, $g \in G'_\infty$, and $k \in \Pi K'_p$. So letting $\{\phi_i\}_{i=1,\cdots,n}$ denote any basis of the subspace of $H'$ taking values in some fixed irreducible subspace of U we use (10.32) and put

$$\theta_{i,j}(g) = \theta_\Phi(d,\pi')(g)$$

substituting $\phi_i$ for $\phi$ and $g_j$ for $g_0$ provided $\det(\phi_i(g_j)) \neq 0$; for $\Phi$ we take $\Pi\Phi_p$ where $\Phi_\infty(x) = e^{2\pi N(x)}tr(\rho_{k-2}(x^\sigma))$ and $\Phi_p$ is the characteristic function of $\mathfrak{D}_p$ for all finite p.

THEOREM 10.13. *Let* $H(\psi,N,k)$ *denote the space of functions* $\phi$ *in* $A_0(\psi)$ *satisfying conditions (ii)-(iv) of (3.6) of Section 3. Let* $H_0(\psi,N,k)$ *denote the intersection of* $H(\psi,N,k)$ *with those subspaces of* $L_0^2(G_Q\backslash G_A,\psi)$ *belonging to any* $\pi$ *as above. Then* $H_0(\psi,N,k)$ *is spanned by the collection* $\{\theta_{i,j}\}$, $i,j = 1,\cdots,n$.

*Proof.* The intersection of $H(\psi,N,k)$ with any of these (finitely many) appropriate $\pi$ is one-dimensional and corresponds to a new form in $S_k(N,\psi)$. Suppose $\pi'$ is the automorphic form on $G'_A$ such that $\pi(\pi') = \pi$ (Theorem 10.5, Part (ii)). Then there is a $(d,\pi')$ "spherical function" $\phi_{\pi'}$ such that $\theta_{\phi,g}$ belongs to the space of $\pi(\pi')$ and so since $\{\phi_{ij}\}$ generates all possible $(d,\pi')$ spherical functions (for these $\pi'$) the theorem follows. □

REMARK 10.14 (*Concerning the classical content of Theorem 10.12*). The theta-functions $\theta_{ij}$ intersect the theta-series defined classically using quaternary quadratic forms and spherical harmonics (these harmonics being the basic spherical functions for SU(2)).

In [Eichler 3] it is proved that $S_k(\Gamma_0(p))$ is spanned by *old forms* for $\Gamma_0(p)$ and theta-series (with spherical harmonics of weight k–2) attached to a definite division quaternion algebra of discriminant p. Keeping in mind that elements of $H(\psi, N, k)$ are *new forms* for $S_k(N, \psi)$ it follows that Theorem 10.12 contains this result.

FURTHER NOTES AND REFERENCES

The arithmetical theory of quaternion algebras is developed in [Eichler] and summarized among other places, in [Eichler 2].

Various Dirichlet series and automorphic forms on quaternion algebras are discussed in [Shimura 3], [Tamagawa], [Godement 4] and [Selberg]. The general theory is described in [Jacquet-Godement] and [Jacquet-Langlands].

Application of the trace formula to automorphic forms on quaternion algebras is the subject matter of [Eichler 3,4], [Shimizu], [Shimizu 2], and [Jacquet-Langlands]. Our exposition is adapted from [Jacquet-Langlands]. Further arithmetical results relating to the trace formula are described in [Mautner 2] and [Langlands 6].

## BIBLIOGRAPHY


[Arthur], *The Selberg trace formula for groups of F-rank one*. Ann. of Math., to appear.

[Arthur 2], *On some tempered distributions on semi-simple groups of real rank one*. Ann. of Math., to appear.

[Artin], *Uker eine neue Art von L-Reihen*. Abh. Math. Sem. Univ. Hamburg 3 (1923).

[Atkin-Lehner], *Hecke operators on* $\Gamma_0(m)$. Math. Ann. 185 (1970), pp. 134-160.

[Bargmann], *Irreducible representations of the Lorentz group*. Ann. of Math., 48 (1947), pp. 568-640.

[Borel], *Some finiteness properties of adele groups over number fields*. Inst. Hautes Etudes Scient. 16 (1963), pp. 5-30.

[Cartier], *Some numerical computations relating to automorphic functions*, in *Computers in Number Theory*. Academic Press, 1971.

[Casselman], *On some results of Atkin and Lehner*. Math. Ann. 201 (1973), pp. 301-314.

[Cassels-Fröhlichs], *Algebraic Number Theory*. Thompson Book Company, Washington D. C., 1967.

[Deligne], *Formes modulaires et representations* $\ell$*-adiques*. Seminaire Bourbaki, 1968/1969, n° 355.

[Deligne 2], La Conjecture de Weil, preprint. I.H.E.S., 1973.

[Duflo-Labesse], *Sur la formule des traces de Selberg*. Ann. Scient. Ecole Norm. Sup., 4e séries, t.4 (1971).

[Eichler], *Zur Zahlentheorie der Quaternionen-Algebren*. Journ. reine angew. Math. 195 (1956), pp. 127-151.

[Eichler 2], *Lectures on Modular Correspondences*. Tata Institute, 1956.

[Eichler 3], *Uber die Darstellberkeit von thetafunktionen durch Thetareihen*. Journ. reine angew. Math., 195 (1956), pp. 156-171.

[Eichler 4], *The basis problem for modular forms and the traces of the Hecke operators*, in "Modular Functions of One Variable I." Springer Lecture Notes, Vol. 320, 1973.

[Gangolli], *Spectra of discrete subgroups*, in *Proceedings* of Symposia in Pure Mathematics, A.M.S. Vol. 26, pp. 431-436.

[Gelbart], *Fourier analysis on matrix space*. Memoirs of the A.M.S., No. 108, 1971.

[Gelbart 2], *An example in the theory of automorphic forms*, in *Proceedings* of Symposia in Pure Mathematics, A.M.S., Vol. 26, pp. 437-439.

[Gelbart 3], *Automorphic forms on the metaplectic group*, in preparation.

[Gelbart 4], *Holomorphic discrete series for the real symplectic group*. Inventiones Math. 19 (1973), pp. 49-58.

[Gelfand-Fomin], *Geodesic flows on manifolds of constant negative curvature*. Translations of the A.M.S., Series 2, Vol. 1 (1965), pp. 49-65; the Russian paper appeared in Uspekhi Mat. Nauk in 1952.

[Gelfand-Graev-Pyatetskii-Shapiro], *Representation Theory and Automorphic Functions*. W. B. Saunders Co., Philadelphia, 1969.

[Godement], *Notes on Jacquet-Langlands' theory*. Institute for Advanced Study, Princeton, 1970.

[Godement 2], *The spectral decomposition of cusp forms*, in *Proceedings* of Symposia in Pure Mathematics, A.M.S., Vol. 9, pp. 225-234. Providence, R. I.

[Godement 3], *Travaux de Hecke* II. Seminaire Bourbaki, 1951/1952, n° 59.

[Godement 4], *Analyse spectrale des functions modulaires*. Seminaire Bourbaki, 1964/65, no. 278.

[Godement 5], *Les functions* $\zeta$ *des algebres simples*, I, II. Seminaire Bourbaki, 1958/1959.

[Godement-Jacquet], *Zeta-functions of simple algebras*. Springer Lecture Notes, Vol. 260, 1972.

[Goldstein], *Analytic Number Theory*. Prentice-Hall, 1971.

[Gross-Kunze], *Bessel transforms and discrete series*, in Springer Lecture Notes, Vol. 276 (1972).

[Gunning], *Lectures on modular forms*. Ann. of Math. Studies, Princeton Univ. Press, Princeton, 1962.

[Harish-Chandra], *Automorphic forms on a semi-simple Lie group*. Proc. Nat. Acad. Sci., U.S.A., 45 (1959), 570-573.

[Harish-Chandra 2], *Representations of semi-simple groups* II. T.A.M.S., Vol. 76, 1954.

[Harish-Chandra 3], *Representation theory of* p-*adic groups*, in *Proceedings* of Symposia in Pure Mathematics. Vol. 25, Providence, R.I.

[Harish-Chandra 4], *Automorphic forms on semi-simple Lie groups*. Springer Lecture Notes, Vol. 62, 1968.

[Harish-Chandra 5], *Eisenstein series over finite fields*, in "Functional Analysis and related fields," pp. 76-88, Springer Verlag, 1970.

[Hecke], *Mathematische Werke*. Göttingen, 1959.

[Howe], *Some qualitative results on the representation theory of* GL(n) *over a* p-*adic field*, preprint (1972).

[Jacquet], *Representations des groupes lineaires* p-*adiques*. C.I.M.E. Summer School, Montecatini (1970).

[Jacquet 2], *Automorphic forms on* GL(2), Part II. Springer Lecture Notes, Vol. 278, 1972.

[Jacquet-Langlands], *Automorphic forms on* GL(2). Springer Lecture Notes, Vol. 114, 1970.

[Knapp-Stein], *Intertwining operators for semi-simple groups*. Ann. of Math. 93 (1971), pp. 489-578.

[Kneser], *Strong approximation*, in *Proceedings* of Symposia in Pure Mathematics, Vol. 9, pp. 187-196. Providence, R. I.

[Kubota], *Elementary Theory of Eisenstein Series*. Kodansha Ltd. and John-Wiley, 1973.

[Kunze-Stein], *Uniformly bounded representations and harmonic analysis of the* 2×2 *real unimodular group*. Amer. J. of Math., Vol. 82 (1960, pp. 1-62.

[Labesse], L-*indistinguishable representations and trace formula for* GL(2), preprint 1972.

[Langlands], *On the functional equations satisfied by Eisenstein series*, mimeographed notes, Yale University.

[Langlands 2], *Problems in the theory of automorphic forms*, in Springer Lecture Notes, No. 170 (1970), pp. 18-86.

[Langlands 3], *Euler products*. Yale Univ. Press, New Haven, 1971; James Whittemore Memorial Lectures 1967.

[Langlands 4], *On the functional equation of the Artin* L-*functions*, mimeographed notes, Yale University.

[Langlands 5], *Eisenstein Series*, in *Proceedings* of Symposia in Pure Mathematics, A.M.S., Vol. 9, pp. 235-257.

[Langlands 6], *A little bit of number theory*, notes 1973.

[Maass], *Über eine neue Art von nicht analytischen automorphen Funktionen und die Bestimmung Dirichletscher Reihen durch Funktionalgleichungen*. Math. Ann. 121 (1949), pp. 141-183.

[Mautner], *Spherical functions over* p-*adic fields*. Amer. J. of Math. Vol. 58, 1958.

[Mautner 2], *Sur certaines formules de trace explicites*, I-III. C. R. Acad. Sc. Paris (1972) Série A, T. 274, pp. 1092-1095, t. 275, pp. 353-356, 739-742.

[Miyake], *On automorphic forms on* $GL_2$ *and Hecke operators*. Ann. of Math., 94 (1971), pp. 174-189.

[Ogg], *Modular forms and Dirichlet series*. Benjamin, Inc., New York, 1969.

[Robert], *Formes automorphes sur* $GL_2$ [*Travaux de H. Jacquet et R. P. Langlands*]. Seminaire Bourbaki, 1971/1972, no. 415.

[Roelcke], *Über die Wellengleichungen bei Grenzkreisgruppen erster Art*, Abh. Heidelberg Akad. Wiss. 4 (1956), pp. 159-267.

[Sally-Shalika], *The Fourier transform on* $SL_2$ *over a non-archimedean local field*, preprint.

[Sally-Warner], *The Fourier transform on semi-simple Lie groups of real rank one*. Acta Math., to appear.

[Satake], *Spherical functions and Ramanujan conjecture*, in *Proceedings* of Symposia in Pure Mathematics. Vol. 9, A.M.S., Providence, R. I., pp. 258-264.

[Selberg], *Harmonic analysis and discontinuous groups in weakly symmetric Riemannian spaces with applications to Dirichlet series*. J. Indian Math. Soc., 20 (1956), pp. 47-87.

[Serre], *A Course in Arithmetic*. Springer-Verlag, 1973.

[Shalika], *Representations of the two by two unimodular group over local fields*, thesis. Johns Hopkins University, 1966.

[Shalika 2], *Some conjectures in class-field theory*, in *Proceedings* of Symposia in Pure Mathematics, A.M.S., Vol. 20, pp. 115-122, Providence, R.I.

[Shalika-Tanaka], *On an explicit construction of a certain class of automorphic forms*. Amer. J. of Math., Vol. 91 (1969), pp. 1049-1076.

[Shimizu], *On zeta functions of quaternion algebras*. Ann. of Math. 81 (1965), pp. 166-193.

[Shimizu 2], *Theta series and automorphic forms on* $GL_2$. Jour. of the Math. Soc. of Jap., 24 (1972), pp. 638-683.

[Shimura], *Introduction to the arithmetic theory of automorphic function*. Iwanami Shoten and Princeton University Press, 1971.

[Shimura 2], *On modular forms of half-integral weight*. Ann. of Math., 97 (1973).

[Shimura 3], *On Dirichlet series and abelian varieties attached to automorphic forms*. Ann. of Math., 76 (1962).

[Silberger], $PGL_2$ *over the* p-*adics*. Springer Lecture Notes, Vol. 166, 1970.

[Stein], *Analysis in matrix spaces and some new representations of* SL(N,C). Ann. of Math., 86 (1967).

[Tamagawa], *On* $\zeta$*-functions of a division algebra*. Ann. of Math., 77 (1963).

[Tanaka], *On irreducible unitary representations of some special linear groups of the second order*, I. Osaka Journal of Math., Vol. 3 (1966), pp. 217-227.

[Weil], *Über die Bestimmung Dirichletscher Richen durch Funktional gleichungen*. Math. Ann. 168 (1967); pp. 149-1ᴗ6.

[Weil 2], *Dirichlet series and automorphic forms*. Springer Lecture Notes, Vol. 189, 1971.

[Weil 3], *Sur certain groupes d'opérateurs unitaires*. Acta. Math. 111 (1964), pp. 143-211.

[Weil 4], *Basic Number Theory*. Springer-Verlag, 1967.

[Weil 5], *Sur la théorie du corps de classes*. J. Math. Soc. Japan 3 (1951), pp. 1-35.

[Weil 6], *Zeta functions and distributions*. Seminaire Bourbaki, No. 312, 1965-66.

Added in proof (September 23, 1974): the reader should now also consult Volumes II and III of the Proceedings of the International Summer School on "Modular Functions of One Variable," Springer Lecture Noιes, Volumes 349 and 350, 1973; especially relevant are the papers of Casselman and Deligne in Volume II.

INDEX

www.ingramcontent.com/pod-product-compliance
Ingram Content Group UK Ltd.
Pitfield, Milton Keynes, MK11 3LW, UK
UKHW042228130125
453571UK00001B/44

9 780691 081564